Philosophical Studies Series

Volume 126

More information about this series at http://www.springer.com/series/6459

Linnet Taylor • Luciano Floridi
Bart van der Sloot
Editors

Group Privacy

New Challenges of Data Technologies

Springer

Editors
Linnet Taylor
Tilburg Institute for Law, Technology
and Society
Tilburg University
Tilburg, The Netherlands

Luciano Floridi
Oxford Internet Institute
University of Oxford
Oxford, UK

Bart van der Sloot
Tilburg Institute for Law, Technology
and Society
Tilburg University
Tilburg, The Netherlands

Philosophical Studies Series
ISBN 978-3-319-46606-4 ISBN 978-3-319-46608-8 (eBook)
DOI 10.1007/978-3-319-46608-8

Library of Congress Control Number: 2016961701

Printed on acid-free paper

This Springer imprint is published by Springer Nature
The registered company is Springer International Publishing AG
The registered company address is: Gewerbestrasse 11, 6330 Cham, Switzerland

Acknowledgements

This book had its genesis in a serendipitous conversation between Linnet Taylor and Luciano Floridi at the Oxford Internet Institute in early 2014. Subsequently, Mireille Hildebrandt became part of this discussion, and in September 2014, in cooperation with Bart van der Sloot, we organised a workshop on the topic of group privacy at the University of Amsterdam which generated several of the chapters that follow. We thank Isa Baud, Karin Pfeffer and the Governance and International Development group at the University of Amsterdam for supporting that workshop and also attendees including Mireille Hildebrandt, Beate Roessler, Nico van Eijk, Julia Hoffman and Nishant Shah, who contributed important ideas and insights to the discussion.

For further illuminating conversations, insights and opportunities, we also thank Julie Cohen, Nicolas de Cordes, Rohan Samarajiva and Gus Hosein.

Finally, with thanks to Dennis and Olivia Broeders for patience and support, and to Julia, who was there right at the start.

Contents

About the Authors

Thomas Baar Within HumanityX (Centre for Innovation, Leiden University), Thomas supports organisations working in the peace, justice and humanitarian sector to spearhead innovations in order to increase their impact on society. As part of an interdisciplinary team, he helps partners to turn ideas into working prototypes over short periods of time. With a background in conflict studies and responsible innovation, he focuses in his work and research on both the opportunities and (data responsibility) challenges offered by data-driven innovations for peace and justice.

Jos Berens was educated in law and philosophy, and having held prior positions at the Dutch Foreign Ministry and the World Economic Forum, currently heads the Secretariat of the International Data Responsibility Group, a collaboration between the Data & Society Research Institute, Data-Pop Alliance, the GovLab at NYU, Leiden University and UN Global Pulse. Together, these partners advance the agenda of responsible use of digital data for vulnerable and crisis-affected populations. Jos is project officer at Leiden University's Centre for Innovation, where he focuses on the risks, and the ethical and legal aspects of projects in the HumanityX program.

Paul de Hert is an international fundamental rights expert. The bulk of his work is devoted, but not limited, to technology & privacy law, criminal law, human rights law and constitutionalism in an historical perspective. In Brussels, Prof. Dr. De Hert holds the chair of "Criminal Law" and "International and European Criminal Law". At the Vrije Universiteit Brussel, he is the Director of the Research group on Fundamental Rights and Constitutionalism (FRC), Director of the Department of Interdisciplinary Studies of Law (Metajuridica), Co-Director of the Research group on Law, Science, Technology & Society (LSTS) and Co-Director of the Brussels Privacy Hub. In Tilburg he holds a position as an associated-professor in the Institute of Law and Technology at the Tilburg University. He is a member of the editorial boards of journals such as the Inter-American and European Human Rights Journal

(Intersentia), Criminal Law & Philosophy (Springer) and The Computer Law & Security Review (Elsevier). He is co-editor in chief of the Supranational Criminal Law Series (Intersentia) and the New Journal of European Criminal Law (Intersentia).

Quirine Eijkman (PhD.) is a senior researcher/lecturer at the Centre for Terrorism and Counterterrorism of the Faculty Campus The Hague, Leiden University, and the head of the Political Affairs and Press Office of Amnesty International Dutch section. This paper is written in her personal capacity. Her research focuses on the (side) effects of security governance for human rights, transitional justice and the sociology of law. She teaches (master) courses on security and the rule of law and international crisis and security management.

Luciano Floridi is professor of philosophy and ethics of information at the University of Oxford, where he is the director of research of the Oxford Internet Institute. Among his recent books, all published by Oxford University Press, are *The Fourth Revolution: How the Infosphere Is Reshaping Human Reality* (2014), *The Ethics of Information* (2013) and *The Philosophy of Information* (2011). He is a member of the EU's Ethics Advisory Group on Ethical Dimensions of Data Protection and of the Google Advisory Board on "the right to be forgotten" and chairman of the Ethics Advisory Board of the European Medical Information Framework.

Dara Hallinan studied law in England and Germany and completed a Master's in Human Rights and Democracy in Italy and Estonia. From 2011, he worked at Fraunhofer ISI before moving to FIZ Karlsruhe – Leibniz-Institut für Informationsinfrastruktur in 2016. The focus of his work is the interaction between new technologies – particularly ICT and biotechnologies – law and society. He is writing his PhD on 'The Role of Data Protection Law in Protecting Genetic Privacy in Research Biobanking' at the Vrije Universiteit Brussel in Belgium.

Lanah Kammourieh is a privacy and cybersecurity lawyer and policy professional. She is also a doctoral candidate at Université Panthéon-Assas (Paris 2). Her legal research has spanned topics in public international law, such as the lawfulness of drones as a weapons delivery platform, as well as privacy law, such as the compared protection of email privacy under U.S. and E.U. legislation. She is a graduate of Université Panthéon-Assas, Sciences Po Paris, Columbia University, and Yale Law School.

Emmanuel Letouzé is the director and co-founder of Data-Pop Alliance. He is a visiting scholar at MIT Media Lab, a fellow at HHI, a senior research associate at ODI, a non-resident adviser at the International Peace Institute and a PhD candidate (ABD) in demography at UC Berkeley. His interests are in Big Data and development, conflict and fragile states, poverty, migration, official statistics and fiscal policy. He is the author of the UN Global Pulse's White Paper, "Big Data for Development:

Challenges and Opportunities", where he worked as senior development economist in 2011–2012, and the lead author of the report "Big Data for Conflict Prevention" and of the 2013 and 2014 OECD Fragile States reports. In 2006–2009 he worked for UNDP in New York, including on the Human Development Report research team. In 2000–2004 he worked in Hanoi, Vietnam, for the French Ministry of Finance as a technical assistant on public finance and official statistics. He is a graduate of Sciences Po, Paris (BA, political science, 1999; MA, economic demography, 2000), and Columbia University (MA, 2006), where he was a Fulbright Fellow.

Julia Manske co-leads the project "Open Data & Privacy" at Stiftung Neue Verantwortung (SNV), a Berlin-based think tank. In this responsibility she works on the development of privacy frameworks for sharing and using data, for instance in smart city contexts. Furthermore, Julia has expertise in digital policies and digital rights in the context of global development. She is a member of Think Tank 30, an offshoot of the Club of Rome, a Research Affiliate with Data-Pop Alliance in New York and is a Global Policy Fellow of ITS in Rio de Janeiro.

Alessandro Mantelero is full-tenured aggregate professor of private law at the Polytechnic University of Turin, director of privacy and faculty fellow at the Nexa Center for Internet and Society and research consultant at the Sino-Italian Research Center for Internet Torts at Nanjing University of Information Science and Technology. Alessandro Mantelero's academic work is primarily in the area of law and technology. His research has explored topics including data protection, legal implications of cloud computing and Big Data, robotics law, internet law, e-government and e-democracy.

Kieron O'Hara is a senior research fellow in electronics and computer science at the University of Southampton, UK, with research interests in trust, privacy and the politics of Web technology. He is the author of several books, including *The Spy in the Coffee Machine: The End of Privacy as We Know It* (2008, with Nigel Shadbolt) and *The Devil's Long Tail: Religious and Other Radicals in the Internet Marketplace* (2015, with David Stevens).

Ugo Pagallo is professor of jurisprudence at the Department of Law, University of Turin, since 2000, faculty at the Center for Transnational Legal Studies (CTLS) in London and faculty fellow at the Nexa Center for Internet and Society at the Polytechnic University of Turin. Member of the European RPAS Steering Group (2011–2012) and the group of experts for the Onlife Initiative set up by the European Commission (2012–2013), he is chief editor of the Digitalica series published by Giappichelli in Turin and coeditor of the AICOL series by Springer. Author of ten monographs and numerous essays in scholarly journals, his main interests are AI and law, network and legal theory, robotics and information technology law (especially data protection law, copyright and online security). He currently is member of the Ethical Committee of the CAPER project, supported by the European Commission through the Seventh Framework Programme for Research and Technological Development.

John Palmer is a Marie Curie Research Fellow and tenure-track faculty member in the Interdisciplinary Research Group on Immigration and the Sociodemography Research Group at Pompeu Fabra University. He works on questions arising in demography, law, and public policy related to human mobility and migration, social segregation, and disease ecology. He has also worked as a protection officer for the U.N. High Commissioner for Refugees in the former Yugoslavia and served as a law clerk, mediator and staff attorney for the U.S. Court of Appeals for the Second Circuit.

Nathaniel Raymond is the director of the Signal Program on Human Security and Technology at the Harvard Humanitarian Initiative (HHI) of the Harvard Chan School of Public Health. He has over 15 years of experience as a humanitarian aid worker and human rights investigator. Raymond was formerly director of operations for the George Clooney-founded Satellite Sentinel Project (SSP) at HHI. Raymond served in multiple roles with Oxfam America and Oxfam International, including in Afghanistan, Sri Lanka, Ethiopia and elsewhere. He has published multiple popular and peer-reviewed articles on human rights, humanitarian issues and technology in publications including the *Georgetown Journal of International Affairs*, *The Lancet*, the *Annals of Internal Medicine* and many others. Raymond served in 2015 as a consultant on early warning to the UN Mission in South Sudan. He was a 2013 PopTech Social Innovation Fellow and is a co-editor of the technology issue of *Genocide Studies and Prevention: An International Journal*. Raymond and his Signal Program colleagues are co-winners of the 2013 USAID/Humanity United Tech Challenge for Mass Atrocity Prevention and the 2012 US Geospatial Intelligence Foundation Industry Intelligence Achievement Award.

Dave Robertson is Professor of Applied Logic and a Dean in College of Science and Engineering at the University of Edinburgh. He is Chair of the UK Computing Research Committee and a member of the EPSRC Strategic Advisory Team for ICT. He is on the management boards for two Scottish Innovation Centres (in Digital Healthcare and in Data Science) and is a member of the Scottish Farr research network for medical data. His current research is on formal methods for coordination and knowledge sharing in distributed, open systems using ubiquitous internet and mobile infrastructures. His current work (on the SociaM EPSRC Programme social.org, Smart Societies European IP smart-society-project.eu and SocialIST coordinating action social-ist.eu) is developing these ideas for social computation. His earlier work was primarily on program synthesis and on the high level specification of programs, where he built some of the earliest systems for automating the construction of large programs from domain specific requirements. He trained as a biologist and remains keen on bio-medical applications, although his methods have also been applied to other areas such as astronomy, healthcare, simulation of consumer behaviour and emergency response.

David Sangokoya is the Research Manager at Data-Pop Alliance and manages the Alliance's research projects on the ethical, governance, and human rights implications of big data. His work focuses on whether and how to leverage big data and algorithms as levers towards inclusive development in developing countries, and additionally, measuring the impact of data "innovations" and data literacy towards greater accountability and agency. Prior to this, David was a research fellow at the Governance Lab, and has previously worked on projects related to postconflict transition, humanitarian-development operations and sustainable development in sub-Saharan Africa and South Asia. He graduated with an MPA in international program management and operations from NYU and a BA with honors in international relations and African studies from Stanford University.

Linnet Taylor is assistant professor of data ethics, law and policy at the Tilburg Institute for Law, Technology, and Society (TILT). She was previously a Marie Curie Research Fellow in the University of Amsterdam's International Development Faculty, with the Governance and Inclusive Development group. Her research focuses on the use of new types of digital data in research and policymaking around issues of development, urban planning and mobility. She was a postdoctoral researcher at the Oxford Internet Institute and studied a DPhil in international development at the Institute of Development Studies, University of Sussex. Her doctoral research focused on the adoption of the Internet in West Africa. Before her doctoral work, she was a researcher at the Rockefeller Foundation where she developed programmes around economic security and human mobility.

Patrick Vinck Ph.D., is the Harvard Humanitarian Initiative's director of research. He is assistant professor at the Harvard Medical School and Harvard T.H. Chan School of Public Health, and lead investigator at the Brigham and Women's Hospital. His current research examines resilience, peacebuilding, and social cohesion in conflicts and disaster settings, as well as the ethics of data and technology in the field. He is the co-founder and director of KoBoToolbox a data collection service, and the Data-Pop Alliance, a Big Data partnership with MIT and ODI.

Bart van der Sloot specialises in questions regarding privacy and Big Data. Funded by a Top Talent grant from the Dutch Organisation for Scientific Research (NWO), his research at the Institute for Information Law (University of Amsterdam) is focused on finding an alternative for the current privacy paradigm, which is focused on individual rights and personal interests. In the past, Bart van der Sloot has worked for the Netherlands Scientific Council for Government Policy (WRR), an independent advisory body for the Dutch government, co-authoring a report on the regulation of Big Data in respect of privacy and security. He currently serves as the general editor of the European Data Protection Law Review and is the coordinator of the Amsterdam Platform for Privacy Research.

Chapter 1
Introduction: A New Perspective on Privacy

Linnet Taylor, Luciano Floridi, and Bart van der Sloot

Abstract The introduction outlines the origin of the book in the concerns arising from new data analytical technologies with regard to collectives. Although their effects have so far been addressed only on the individual level, in fact profiling and machine learning technologies are directed at the group level and are used to formulate types, not tokens – they work to scale, and enable their users to target the collective as much as the individual. This means that our legal, philosophical and analytic attention to the individual may need to be adjusted, and possibly extended, in order to pay attention to the actual technological landscape unfolding before us. This book represents such an adjustment: it may be seen as an exploration of the territory that lies between 'their privacy' and 'its privacy', with regard to a given group. In this book we push the boundary towards 'its', and begin to think about the implications of that shift, first by identifying who must be involved in the discussion. This chapter outlines the philosophical and legal concerns that arise from considering group privacy, and sets out the challenge that this volume aims to answer.

Keywords Collectives • Group privacy • Right to privacy • Relational privacy • Network effects • GDPR • Profiling • Algorithms

L. Taylor (✉) • B. van der Sloot
Tilburg Institute for Law, Technology and Society, Tilburg University, 90153, Warandelaan 2, 5000 LE Tilburg, The Netherlands
e-mail: l.e.m.taylor@uvt.nl; B.vdrSloot@uvt.nl

L. Floridi
Oxford Internet Institute, University of Oxford, 1 St Giles, OX1 3JS, Oxford, UK
e-mail: luciano.floridi@oii.ox.ac.uk

L. Taylor et al. (eds.), *Group Privacy*, Philosophical Studies Series 126,
DOI 10.1007/978-3-319-46608-8_1

1.1 The Project and Its Origins

This book is the product of an interdisciplinary discussion that began from a single observation: that group privacy seems to be falling short with regard to emerging data analytic techniques. All around us, data analytic technologies are focused on our lives and our behaviour. Their gaze is rarely focused on individuals, but on the crowd of technology users, a crowd that is increasingly global. Much attention is paid to the concepts of anonymisation, of protecting individual identity, and of safeguarding personal information. However, in an era of big data where analytics are being developed to operate at as broad a scale as possible, the individual is often incidental to the analysis. Instead, data analytical technologies are directed at the group level. They are used to formulate types, not tokens (Floridi, Chap. 5, this volume) and the kinds of actions and interventions they facilitate are aimed beyond individuals. This is precisely the value of big data: it enables the analyst to gain a broader view, to strive towards the universal. Yet even if data analytics do not involve 'piercing the collective shell' (Samarajiva and Lokanathan 2016), they may still result in decisions that pose real risks on the aggregate level, for groups of – or rather grouped – people.

What does this mean for privacy? One implication is that our legal, philosophical and analytic attention to the individual may need to be adjusted, and possibly extended, in order to pay attention to the actual technological landscape unfolding before us. That landscape is one where risks relating to the use of big data may play out on the collective level, and where personal data is at one end of a long spectrum of targets that may need consideration and protection. Taking this as our starting point for this volume, we aim to raise new – and hopefully inconvenient – questions with regard to current conceptualisations of privacy and data protection. One starting point for the project was that the group had not been conceptualised in terms of privacy beyond a collection of individuals with individual interests in privacy (Bloustein 1978). Our central question is whether, and how, we may be able to move from 'their' to 'its' privacy with regard to the group.

Answering this question requires first that we have an idea what kind of group we mean. The authors in this volume offer different perspectives as to the kinds of grouping relevant to privacy and big data: political collectives, groupings created by algorithms, and ethnic groupings are just some of the typologies explored. Some of the groupings dealt with by the contributors are defined by a common threat of harm, some by a similar reason for an interest in privacy, and some by a similar type of privacy interest. This lack of consensus is partly a function of the multidisciplinary nature of the project, since legal scholars will think differently about groups from philosophers, and philosophers differently from social scientists. Given the inadequacy of current approaches to privacy in the face of big data (Barocas and Nissenbaum 2014; Floridi 2014) it is not dogmatism but an expert-led and exploratory debate that may help us to question and move beyond the limitations of current definitions.

Given this exploratory objective, we present a multidisciplinary perspective both in order to highlight the complexity of discussing issues of privacy and data protection across a number of fields where they are relevant concerns, and in order to suggest that the way such a discussion can proceed is by focusing on the data technologies themselves and the problems they present, rather than on the different disciplinary traditions and perspectives involved in the research fields implicated by those technologies. Our approach to defining group privacy aims to be functional and iterative rather than stable and unanimous: it involves a conversation amongst authors from a range of fields that are each faced with this emerging problem, and each of whom may have a piece of the answer.

The fields include legal philosophy, information ethics, human rights, computer science, sociology, and geography. The case studies used include satellite data from Africa, the human genome, and social networks that act as machines. What brings them together is that they deal with types of data that largely did not exist a generation ago, such as genomic information, digital social networks, and mobile phone traces; and with the methods of analysis that are evolving to fit them, such as distributed and cloud computing, machine learning, and algorithmic decision-making. Although several of these are not new, the challenges we address here arise from their use on unprecedentedly large and detailed data or new objects of analysis.

1.2 Emerging Data Technologies and Practices

The new data technologies that are the focus of this book range from the myriad tools and applications available in high-income countries to emerging technologies and uses common in lower-income ones, and from highly networked and monitored environments to those where connectivity is fairly new and awareness of monitoring and profiling is low. Around the world, digitisation and datafication (the transformation of all kinds of information into machine readable, mergeable and linkable form) are providing new sources of data and new analytical possibilities. At the time of writing there are 7.4 billion mobile connections worldwide, 5.5 billion of them in low- and middle-income countries (LMICs), where 2.1 billion people are already online (ITU 2015). LMICs, in fact, have been forecast to provide the majority of geolocated digital data by 2020 (Manyika et al. 2011).

'The god's eye view' that big data provides (Pentland 2011) stems primarily from people's use of digital technology: it is behavioural, granular data that may be de-identified and subjected to a range of aggregation or blurring techniques in terms of individual identity, but still reflects on one level or another the behaviour and activities of those users. This type of data is born-digital, often emitted as a result of activities or transactions, and often where the technology user is not aware of creating those signals and records. The activities include using digital communications technologies such as mobile phones and the internet, conducting transactions using a credit card or a website, being picked up by sensors at a distance such as satellites or CCTV, or by the sensors embedded in the objects and structures we interact with (also known as ubiquitous computing or the Internet of Things). New datasets can also be created by systems that process, link and merge such data, allowing profiles to be constructed that tell the analyst more about the propensities of people or groups.

The emergence of geo-information, the spatial dimension of the data emitted by new digital technologies, is also worth considering as it provides another facet to the possibilities for monitoring, profiling and tracking presence and behaviour. Smartphones in particular are changing the way spatial patterns of people's movements and location can be visualised and monitored, offering signals from GPS, cell tower or wifi connections, Bluetooth sensors, IP addresses and network environment data, all of which can provide a continuous stream of information about the user's activities and behaviour. Geo-information is becoming essential to the 40-billion-dollar global data market because it allows commercial data analysts to distinguish between a human and a bot – an entity that is created to generate content and responses on social media and shows what look like activities, but is not human. From a commercial perspective, a geo-spatial signature on online activity adds value for advertisers and marketers (some of the chief actors in profiling) because location and movement traces guarantee the online presence is human. Apple shares geo-information from its devices commercially; 65.5 billion geotagged payments are made per year in the US alone, and companies such as Skyhook Wireless pinpoint millions of users' WiFi locations daily across North America, Europe, Asia, and Australia (de Montjoye et al. 2013).

The uses of the 'god's eye view' are myriad. The new data sources facilitate monitoring and surveillance, either directed toward care (human rights, epidemiology, 'nowcasting' of economic trends or shocks) or control (security, anti-terrorism) (Lyon 2008). They also allow sorting and categorising ranging from the profiling of possible security threats or dissident activists to biometrics and welfare delivery systems and poverty mapping in lower-income countries. They can be used to identify trends, for example in the fields of economics, human mobility, urbanisation or health, or to understand phenomena such as the genetic origins of disease, migration trajectories, and resource flows of all kinds. The new data sources also allow authorities (and others, including researchers and commercial interests [Taylor 2016]) to influence and intervene in situations ranging from everyday urban or national governance to crisis response and international development. Influencing, profiling, nudging and otherwise changing behaviour is one of the chief reasons big data is generating interest across sectors: from basic research to policy, politics and commerce, the new data sources are being conceptualised as tools that may revolutionise practices of persuading and influencing as much as those of analysing and understanding. The scale of the data, however, means that influence (and the analysis and understanding that facilitate it) is as likely to take place on the demographic as the individual level, and to be conceptualised as moving the crowd as much as changing micro-level patterns of behaviour.

1.3 Transcending the Individual

The search for group privacy can be explained in part by the fact that with big data analyses, the particular and the individual is no longer central. In these types of processes, data is no longer gathered about one specific individual or a small group of people, but rather about large and undefined groups. Data is analysed on the basis of patterns and group profiles; the results are often used for general policies and applied on a large scale. The fact that the individual is often no longer central, but incidental to these types of processes, challenges the very foundations of most currently existing legal, ethical and social practices and theories. The technological possibilities and the financial costs involved in data gathering and processing have for a long time limited the amount of data that could be gathered, stored, processed and used. Because of this limitation, choices had to be made regarding which data was gathered, about which person, object or process, and how long it would be stored. Because data processing consequently often affected only individuals or smaller groups, the social, legal and ethical norms that were developed focussed on the individual, on the particular. Capacities for data processing have grown over the years and the costs have decreased incrementally, and the increasingly large amounts of data processed are still developing exponentially. Big data analytics and the possibilities they offer for gathering, analysing and using huge amounts of data, however, seem to bring not only a quantitative, but also a qualitative shift. They challenge the fundamental basis of the social, legal and ethical practices and theories that have been developed and applied over decades.

As is noted by a number of authors in this book, the current guidelines for data processing are based on personally identifying information. For example, the OECD guidelines define personal data as any information relating to an identified or identifiable individual; the EU Data Protection Directive adds that an identifiable person is one who can be identified, directly or indirectly, in particular by reference to an identification number or to one or more factors specific to his or her physical, physiological, mental, economic, cultural or social identity. Other instruments may use slightly different terminology, but what all of them share is the focus on the individual and the ability to link data back to a particular person or to say something about that person on the basis of the data. Although this focus on personally identifying information is still useful for more traditional data processing activities, it is suggested by many that in the big data era, it should be supplemented by a focus on identifying information about categories or groups.

Since the currently dominant social, legal and ethical paradigms focus primarily on individual interests and personal harm, privacy and data protection are said to be individual interests, either protecting a person's individual autonomy, human dignity, personal freedom or interests related to personal development and identity. Consequently, the assessment of whether a data processing activity does harm or good (identified as the 'non-maleficence' and the 'benevolence' principles by Raymond in this book), is done on the level of the individual, of the particular. However, although specific individuals may be harmed or benefited by certain data uses, this again is increasingly incidental in the big data era. Policies and decisions are made on the basis of profiles and patterns and as such negatively or positively affect groups or categories. This is why it has been suggested that the focus should be on group interests: whether the group flourishes, whether it can act autonomously, whether it is treated with dignity. The harm principle as well as the benevolence principle could subsequently be translated to a higher (non-particular) level as well.

As a final example, the current paradigm focusses on individual control over personal data. The notion of 'informed consent', deeply embedded in Anglo-Saxon thinking about data processing, for example, spells out that personal data may in principle only be gathered, analysed and used if the data subject has consented to it, the consent being specific, freely given and based on full and adequate information. Although in continental European data protection instruments the notion of 'informed consent' plays a limited role, they do give the individual a right to access, correct, control and delete data. The question, however, is whether this focus on individual control still holds in the big data era; given the sheer amount of data processing activities and the size of databases, it becomes increasingly difficult for an individual to be aware of every data processing activity that might include their data, to assess in how far the processing is done legitimately and if not, to request the data controller to stop their activities or ultimately to go to a judge.

The basic agreement amongst most contributors to this book is consequently that the focus on the individual, personal data, individual interests and informed consent or individual control over data is too narrow and should be supplemented by an interpretation of privacy which takes account of broader data uses, interests and practices. In the search for theories in which the focus on the individual is transcended, we have coined the term 'group privacy', though in reality authors differ in their terminology, categorization and solutions to a large extent. Nevertheless, this books tries to determine the basis for conceptualizing the idea of group privacy and to bring the discussion on it to a higher level.

1.4 Conceptualising Group Privacy

One major difficulty in discussing group privacy is representing the nature of the entity in question. A common view is that one may have to identify groups first, in order to be able to discuss properties of such entities, including their potential rights, and hence privacy. It is a set-theoretic, implicit assumption, according to which one has to identify "things" first (these are known as constants or variables and are the bearers of the properties, the elements of the set) and then their properties (known as predicates, or relations). After that, any quantification concerns the "things" (the elements of the set), with "any", "some", "none" or "all" indicating which groups do or do not enjoy a particular property (predicate). This approach is not mistaken in general, but in this case it is most unhelpful because it generates an unnecessary difficulty. Groups are usually dynamic entities: they come in an endless number of sizes, compositions, and natures, and they are fluid. The group of people on the same bus dissolves and recomposes itself at every stop, for example. Fixing them well enough to be able to predicate some stable properties may be impossible. But with groups acting as moving targets and no clear or fixed ontology for them there is little hope a theory of group privacy may ever develop. As a result – the argument concludes – the only fixed entity is actually the individual, so group privacy is nothing more than the sum of privacies enjoyed by the individuals constituting the group. The problem with this line of reasoning is that groups are not "given". Even when they seem to be given – e.g. an ethnic or biological group – it is the choice of a particular property that determines who belongs to that group. It is the property of being "quadrilateral" that puts some figures of the plane in a particular set. Change the property – quadrilateral and rightangled – and the size (cardinality) and composition of the group follows. So a much better alternative is to realise that predicates come first, that groups are constructed according to them, and that, in the case of privacy, it is the same digital technologies used to create a group by selecting some properties rather than others (e.g. "Muslim" instead of "Christian") that can also infringe its privacy. Technologies actually determine groups, through their clustering and typification.

Sometimes such groups overlap with how we group people anyway, e.g. teenagers vs. retired people. Yet this is merely distracting. We are still adopting predicates first. It is just that some of these predicates appear so intuitive as to give us the impression that we are merely describing how the world is, instead of carving it into a shape we then find obvious. So it is misleading to think of a group privacy infringement as something that happens to a group that exists before and independently of the technology that created it as a group. It is more useful to think of algorithms, big data, digital technologies in general as well as information management practices, strategies and policies as designing groups in the first place. They do so by choosing the salient features of interest, according to some particular purpose. This explains why groups are so dynamic: if you change the purpose, you change the set of relevant properties (what in computer science is called the level of abstraction), and obtain a different set of individuals. If what interests you are all the children on the bus because they may need to be accompanied by an adult you obtain a very different outcome than if you are looking for retired people, who may be subject to a discount. To put it simply: the activity of grouping comes before its outcome, the group. This different approach helps to explain why profiling – a standard kind of grouping – may already infringe the privacy of the resulting group, if profiling is oriented by a goal that in itself is not meant to respect the privacy of the group. It also clarifies why group privacy may be infringed even in cases in which the members of the group are not aware of this: a group that has been silently profiled and that is being targeted as a group does not need to know any of this to have a right to see its privacy restored and respected.

If we now return to the previous reasoning about a stable ontology, in the following chapters the reader will encounter two kinds of ontologies. One privileges an individual-based, entity-first approach. When this favours group privacy it tends to do so in a "their" privacy way. If there is such thing as group privacy it is to be analysed as the result of the collection of the privacies of the constituting members. This is like arguing that the set is blue because all its members are blue. The other ontology privileges a property-based, predicate-first approach. When this favours group privacy it tends to do so in a "its" privacy way. If there is such thing as group privacy it is to be analysed as an emergent property, over and above the collection of the privacies of the constituting members. This is like arguing that the set is heavy despite the fact that all its members are light, because many light entities make up a heavy sum.

1.5 The Legal Field's Engagement with Group Privacy

The position of the group in the legal context has been a complex one. It has been argued by some that group rights are the origin of the legal regime as such, or at least of the human rights framework. One of the first fundamental rights to be generally acknowledged was the freedom of religion. This fundamental right was granted in countries in which a majority adhered to one religion, for example the Catholic faith, and a substantial minority adhered to another religion, for example Protestantism. In essence, thus, a group, in this case the Protestants, was granted a liberty through the right to freedom of religion. More in abstract, fundamental rights have always served as counter balance for democracy. While the majority may hold certain beliefs, feel that certain acts should abolished or expressions prohibited, fundamental rights have always guaranteed a minimum amount of freedom, whatever the democratic legislator may enact. That is why fundamental rights have also been called minority rights per se, because they limit the capacity of the majority.

Likewise, with the first real codification of human rights in international law just after the Second World War, the focus was on groups. During that epoch, the fascist regimes, and to a lesser extent the Communist dictatorships, had denied the most basic liberties of groups such as Jews, Gypsies, gays, bourgeoises and intellectuals. The first human rights documents, such as the Universal Declaration of Human Rights (UDHR), the International Covenant on Civil and Political Rights (ICCPR) and the European Convention on Human Rights (ECHR), were all a reaction to the atrocities of the previous decades. They were primarily seen as documents laying down minimum freedoms, liberties which the (democratic) legislator could never curtail, irrespective of whether it concerned the liberties of individuals, groups or even legal persons. For example, in the ECHR not only individuals, but also legal persons and states may complain of a violation of the human rights guaranteed under the Convention, as may groups of natural persons too. The main idea behind these documents was not one of granting subjective rights to natural persons, but rather of laying down minimum obligations for the use of power by states. Consequently, states, legal persons, groups and natural persons could complain if the state exceeded its legal discretion.

However, gradually, this broad focus has been moved to the background in most human rights frameworks, most notably under the European Convention on Human

Rights. The focus has been increasingly on the individual, his or her rights and his or her interests. States seldom file complaints under the ECHR, groups are prohibited from doing so by the European Court of Human Rights (ECtHR) and legal persons are discouraged from submitting complaints, especially under Article 8 of the Convention, containing the right to private life, family life, home and communication. The Court, for a long time, has held as a rule that legal persons cannot complain of a violation of their right to privacy, because according to the ECtHR privacy is so intrinsically linked to individual values that in principle, only natural persons can complain about a violation of this right. Although since 2002, the ECtHR has allowed legal persons to invoke the right to privacy under particular circumstances, these cases are still the exception – in only some ten cases have legal persons been allowed to invoke the right to privacy, very few when compared to the thousands of complaints by natural persons.

Still, there have been some new developments, in particular the idea of third generation rights, minority rights and future generation rights. The right to the respect for minority identity and the protection of the minority lifestyle are partially accepted under the recent case law of the Court, and are commonly considered as rights of groups, such as minorities and indigenous people. These group rights are so-called 'third-generation' rights, which go beyond the scope of the first-generation rights, classic civil and political rights, and socioeconomic rights, which are referred to as second-generation rights, which are mostly characterized as individual rights (Vasak). Third-generation rights focus on solidarity and respect in international, interracial and intergenerational relations. Beside minority rights, third-generation rights include the right to peace, the right to participation in cultural heritage and the right to live in a clean and healthy living environment.

Finally, in privacy literature, the idea of group privacy is not absent (Westin 1967). So-called 'relational privacy' or 'family privacy' is sometimes seen as a group privacy right, at least by Bloustein. However, this right, also protected under the European Convention on Human Rights Article 8, grants an individual natural person the right to protection of a specific interest, namely his interest to engage in relationships and develop family ties – it does not grant a group or a family unit a right to protect a certain group or unit. Attention is also drawn to the fact that the loss of privacy of one individual may have an impact on the privacy of others (Roessler and Mokrosinska 2013). This is commonly referred to as the network effect. A classic example is a photograph taken at a rather wild party. Although the central figure in the photograph may consent to posting the picture of him at this party on Facebook, it may also reveal others attending the party too. This is the case with much information – a person's living condition and the value of his home does not only disclose something about them, but also about their spouse and possibly their children. Perhaps the most poignant example is that of hereditary diseases. In connection to this, reference can be made to the upcoming General Data Protection Regulation, which will likely include rules on 'genetic data', 'biometric data' and 'data concerning health'. Especially, genetic data often tell a story not only about specific individuals, but also about their families or specific family members (see Hallinan & De Hert in this book).

There has always been a troubled marriage between privacy and personality rights. Perhaps one of the first to make a sharp distinction between these two types of rights was Stig Strömholm in 1967 in 'Rights of privacy and rights of the personality: a comparative survey'. He suggested that the right to privacy was a predominantly American concept, coined first by Cooley and made famous by Warren and Brandeis' article 'The right to privacy' from 1890. Personality rights were the key notion used in the European context, having a long history in the legal systems of countries like Germany and France. Although a large overlap exists between the two types of rights, Stömholm suggested that there were also important differences. In short, the right to privacy is primarily conceived as a negative right, which protects a person's right to be let alone, while personality rights also include a person's right to represent his or her self in a public context and develop his or her identity and personality.[1] Although the right to privacy was originally seen as a negative right, the ECtHR has gradually interpreted Article 8 ECHR as a personality right, providing positive freedom to the European citizens and positive obligations for states. The key notion for determining whether a case falls under the scope of Article 8 ECHR seems simply whether a person is affected in his or her identity, personality or desire to flourish to the fullest extent. This practice has had as a consequence that the material scope of the right to privacy has been extended considerably.

The European courts' decisions treat identity and identification as contextual and socially embedded, and consequently as being expressed, asserted or resisted in relation to particular social, economic, or political groupings. The new data technologies, however, pose the question of how people may assert or resist identification when it does not focus on them individually. Although digital technologies have already evolved to be able to identify almost anyone with amazing degrees of accuracy, the fact is that for millions of people this is not relevant. It is often much more valuable – commercially, politically, socially – not to concentrate on an individual – a token – but on many individuals, i.e. the group, clustered by some interesting property – the type to which the token now belongs. Tailoring products or services, for example, means being able to classify tokens like Alice, Bob, and Carol, under the correct sort of type: a skier, a dog lover, a bank manager. "People who bought this also bought …": the more accurate the types, the better the targeting. This is why we shall see a rise in the algorithmic management of data. The more data can be analysed automatically and smartly in increasingly short amounts of time, the more grouping, understood as profiling or as typifying tokens, can become dynamically accurate in real time (Alice does not ski anymore, Bob has replaced his dog with a cat, Carol is now an insurance manager). As algorithmic societies develop, attention to group privacy will have to increase if we wish to avoid abuses and misuses.

The problems of increasingly accurate data are balanced by unpredictabilities and inaccuracies due to the material ways in which communications technologies are accessed and used. For example, in low-income communities multiple people

[1] Elements of this section have been taken from: B. van der Sloot, 'Privacy as personality right'.

may rely on a single mobile phone, meaning that a single data-analytic profile may actually reflect an unknown number of people's activity. Conversely, in areas with poor infrastructure one person may have multiple devices and SIM cards in order to maximise their chances of picking up a signal, which effectively makes them a group for the purposes of profiling (Taylor 2015).

These practices have similar effects to obfuscation-based approaches to privacy (Brunton and Nissenbaum 2013), and therefore have the potential to deflect interventions that rely on accurate profiling. They also, however, may impact negatively on people when that profiling determines important practical judgements about them such as their creditworthiness (is this a group of collaborators suitable for a microfinance intervention, or an individual managing a successful business?), or their level of security threat (is this a network of political dissidents or one person searching for information on security?). Exactly this problem is posed by an experimental credit-rating practice in China which gives firms access to records of people's online activities and those of their friends as a metric for creditworthiness and insurability, and likely soon other characteristics such as visa eligibility and security risk level (Financial Times 2016). The evolution toward systems that rely on granular, born-digital data to categorise people in ways that affect their opportunities and life chances relies heavily on the assumption that individual identities can be mapped directly onto various datafied markers such as search activity, logins and IP addresses. Yet it is clear that individual and group identities bear a complex and highly contextual relationship to each other on both the philosophical and the practical level.

1.6 Conclusion: From 'Their Privacy' to 'Its Privacy'

This book is a conversation that tugs the idea of group privacy in many different directions. It does not aim to be the final answer to what, after all, is an emergent problem, but may be seen as an exploration of the territory that lies between 'their privacy' and 'its privacy', with regard to a given group. By placing the various empirical and legal arguments in dialogue with each other we can push the boundary towards 'its', and by extension, begin to think about the implications of that shift, and identify who must be involved in the discussion in order to best illuminate and address them.

Digital technologies have made us upgrade our views on many social and ethical issues. It seems that, after having expanded our concerns from physical to informational privacy, they are now inviting us to be more inclusive about the sort of entities whose informational privacy we may need to protect. A full understanding of group privacy will be required to ensure that our ethical and legal thinking can address the challenges of our time. We hope this book contributes to the necessary conceptual work that lies ahead.

Bibliography

Barocas, S., and H. Nissenbaum. 2014. Big data's end run around anonymity and consent. In: *Privacy, big data, and the public good: Frameworks for engagement*, 44–75. Cambridge: Cambridge University Press.

Bloustein, E.J. 1978. *Individual and group privacy*. New Brunswick: Transaction Publishers.

Brunton, F., and H. Nissenbaum. 2013. Political and ethical perspectives on data obfuscation. In: *Privacy, due process and the computational turn: The philosophy of law meets the philosophy of technology*, 164–188. New York: Routledge.

de Montjoye, Y.A., C.A. Hidalgo, M. Verleysen, and V.D. Blondel. 2013. Unique in the crowd: The privacy bounds of human mobility. *Scientific Reports* 3: 1376.

Financial Times. 2016. *When big data meets big brother*. January 19, 2016. Accessed 21 Jan 2016 at http://www.ft.com/cms/s/0/b5b13a5eb84711e5b1518e15c9a029fb.html.

Floridi, L. 2014. Open data, data protection, and group privacy. *Philosophy and Technology* 27: 1–3. doi:10.1007/s1334701401578.

ITU. 2015. Key *ICT indicators for developed and developing countries and the world (totals and penetration rates)*. Retrieved from http://www.itu.int/en/ITU-D/Statistics/Pages/stat/default.aspx.

Lyon, D. 2008. *Surveillance society*. Presented at Festival del Diritto, Piacenza, Italia: September 28 2008.

Manyika, J., M. Chui, B. Brown, J. Bughin, R. Dobbs, C. Roxburgh, and A. Hung Byers. 2011. *Big data: The next frontier for innovation, competition and productivity*. Washington, DC: McKinsey Global Institute.

Pentland, A. 2011. Society's nervous system: Building effective government, energy, and public health systems. *Pervasive and Mobile Computing* 7(6): 643–665.

Roessler, B., and D. Mokrosinska. 2013. Privacy and social interaction. *Philosophy & Social Criticism*. July 2013., 0191453713494968.

Samarajiva, R., and S. Lokanathan. 2016. *Using behavioral big data for public purposes: Exploring frontier issues of an emerging policy arena*. LirneAsia report. Retrieved from http://lirneasia.net/wp-content/uploads/2013/09/NVF-LIRNEasia-report-v8-160201.pdf.

Taylor, L. 2015. No place to hide? The ethics and analytics of tracking mobility using mobile phone data. *Environment & Planning D: Society & Space* 34(2): 319–336. doi:10.1177/0263775815608851.

Vasak, K. 1977. 'Human rights: A thirty year struggle: The sustained efforts to give force of law to the universal declaration of human rights', UNESCO Courier 30:11, Paris, United Nations Educational, Scientific, and Cultural Organization.

Westin, A. 1967. *Privacy and freedom*. New York: Atheneum.

Chapter 2
Safety in Numbers? Group Privacy and Big Data Analytics in the Developing World

Linnet Taylor

Abstract This chapter argues that group privacy is a necessary element of a global perpective on privacy. Addressing the problem as a new epistemological phenomenon generated by big data analytics, it addresses three main questions: first, is this a privacy or a data protection problem, and what does this say about the way it may be addressed? Second, by resolving the problem of individual identifiability, do we resolve that of groups? And last, is a solution to this problem transferable, or do different places need different approaches? Focusing on cases drawn mainly from low- and middle-income countries, this chapter uses the issues of human mobility, disease tracking and drone data to demonstrate the tendency of big data to flow across categories and uses, its long half-life as it is shared and reused, and how these characteristics pose particular problems with regard to analysis on the aggregate level.

Keywords Drones • Epidemiology • Migration • Ebola • Mapping • Satellites • Mobile phones • Kenya • Sudan • Africa • Data mining • Predictive modelling

2.1 Introduction

As a way of keeping track of human behaviour and activities, big data is different from previous methods. Traditionally, gathering population data has involved surveys conducted on the individual level with people who knew they were offering up personal information to the government. The census is carefully guarded by the public authorities, and misuse of its data is trackable and punishable. Big data, in contrast, is kept largely by corporate guardians who promise individuals anonymity in return for the use of their data. As Barocas and Nissenbaum (2014) and Strandburg (2014) have shown, however, this promise is likely to be broken because, although big data analytics may allow the individual to hide within the crowd, they cannot

L. Taylor (✉)
Tilburg Institute for Law, Technology and Society, Tilburg University
90153, Warandelaan 2, 5000 LE Tilburg, The Netherlands
e-mail: l.e.m.taylor@uvt.nl

L. Taylor et al. (eds.), *Group Privacy*, Philosophical Studies Series 126,
DOI 10.1007/978-3-319-46608-8_2

conceal the crowd itself. We may be profiled in actionable ways without being personally identified. Thus the way that current understandings of privacy and data protection focus on individual identifiability becomes problematic when the aim of an adversary is not to identify individuals, but to locate a group of interest – for example an ethnic minority, a political network or a group engaged in particular economic activities.

This chapter will explore whether the problems raised by aggregate-level conclusions produced from big data are different from those that arise when individuals are made identifiable. It will address three main questions: first, is this a privacy or a data protection problem, and what does this say about the way it may be addressed? Second, by resolving the problem of individual identifiability, do we resolve that of groups? And last, is a solution to this problem transferable, or do different places need different approaches? To answer these questions, this chapter will focus mainly on data originating outside the high-income countries where debates on privacy and data protection are currently taking place. Looking at three cases drawn mainly from the developing world, I will demonstrate the tendency of big data to flow across categories and uses, its long half-life as it is shared and reused, and how these characteristics pose particular problems with regard to analysis on the aggregate level.

I will argue that in this context, there is no safety in numbers. If groupings created through algorithms or models expose the crowd to influence and possible harm, the instruments that have been developed to protect individuals from the misuse of their data are not helpful. This is for several reasons: first, because when misuse occurs on the group level, individuals remain anonymous and there is no obligation to inform them that their data is being processed. Second, because it is virtually impossible for anyone to know if a particular individual has been subjected to data misuse, a problem not visualised by existing forms of data protection. And third, because many of the uses of big data that involve algorithmic groupings are covered by exceptions to the data protection rules: they are for purposes of scientific research, national security, defence, public safety, or important economic or financial interests on the national level. In the case of many lower-income countries,[1] most data processing is covered either by no data protection legislation at all (Greenleaf 2013) or by legislation that is unenforceable since the processing occurs within multinational companies not situated in the country in question (Taylor 2016b).

What does 'the group' mean? I deal here with groups not as collections of individual rights (Bloustein 1978) but as a new epistemological phenomenon generated by big data analytics. The groups created by profiling using large datasets are different from conventional ideas of what constitutes a group in that they are not

[1] LMICs here are defined according to the World Bank's definitions grouping countries, see: http://data.worldbank.org/about/country-classifications, where LMICs have incomes of US$1036 – $12,616 per capita and high income countries (HICS) above that threshold. My particular focus is the low- and lower-middle-income countries, with an upper threshold of $4085 per capita, which includes India and most of Africa.

self-constituted but grouped algorithmically, and the aim of the grouping may not be to access or identify individuals. Such groupings are practically fuzzy, since they do not focus on individuals within the group, but epistemologically precise because they create a situation where people effectively self-select for a particular intervention due to certain preferences or characteristics. For example, in the Netherlands the city of Eindhoven's Living Lab project exposes people who spend time in particular areas at night under particular conditions (busy streets, many people visiting bars and night-clubs) to behaviour-altering scents, lights and colours (Eindhoven News 2014). In this situation, people self-select into the intervention by going out in the centre of town at night, but are not targeted due to any particular aspect of their individual identity other than their presence in a particular place at a particular time.

Although the implications of data-driven profiling have been analysed in detail across a range of research disciplines (notably in Hildebrandt and Gutwirth 2008), new applications of data technologies are emerging that blur the definition of targeting. In the example of Eindhoven, the intervention cannot be classified as resulting from 'indirect profiling' as defined by Jacquet-Chiffelle (2007:40), which 'aims at applying profiles deduced from other data subjects to an end user', but is instead aimed at all of those who share a particular spatial characteristic (their location) plus a particular activity (visiting bars or clubs in a given area). People are not aware they are being grouped in this way for an intervention, just as people using mobile phones are not aware that researchers may be categorising them into clusters through the analysis of their calling data (e.g. Caughlin et al. 2013). Therefore one central characteristic of the type of grouping this chapter addresses is that of being defined remotely by processing data, so that the group's members are not necessarily aware that they belong to it.

These types of algorithmic, rather than self-constituted, groupings illuminate the problems that can arise from the analysis of deidentified data, and suggest the need to address problems of the group with regard to risk and protection. One is that today, these cluster-type groupings are a source of information for making policy decisions. Another reason is that being able to find groups through their anonymous digital traces offers opportunities to oppressive or authoritarian powers to harm the group or suppress its activities. Increasingly policymakers are looking to big-data analytics to guide decision-making about everything from urban design (Bettencourt 2014) to national security (Lyon 2014). This is particularly the case where developing countries (referred to hereafter as Low and Middle-Income Countries, or LMICs) are concerned. Statistical data for these countries has traditionally been relatively poor (Jerven 2013), so that policymakers are seeking new data sources and analytical strategies to define the target populations for development interventions such as health (Wesolowski et al. 2012), disaster response (Bengtsson et al. 2011) and economic development (Mao et al. 2013). Big data analytics, and mobile phone traces in particular, are the prime focus of this search (World Economic Forum 2014).

Barocas and Nissenbaum (2014) have pointed out how the era of big data may pose new questions to do with privacy on the group level, in contrast to the individual level on which it has traditionally been conceptualised. They argue that big data is different from single digital datasets because it is used in aggregated form, where harm is less likely to be caused by access to personally identifiable information on individuals and more likely to occur where authorities or corporations draw inferences about people on the group level. Their conceptualisation of the problem suggests that if it is to remain relevant, the idea of privacy must be stretched and reshaped to help us think about the group as well as the individual – just as it has been stretched and reshaped beyond Brandeis' original framing as 'the right to be left alone' to cover issues such as intellectual freedom and the right not to be subjected to surveillance (Richards 2013). In particular, the idea of privacy must extend to cover the new types of identifiability occurring due to datafication (Strandburg 2014) in LMICs, which may create or exacerbate power inequalities and information asymmetries.

The cases outlined in this chapter centre around new and emerging uses of digital data for profiling groups that are occurring or being developed worldwide. They are chosen because they involve complementary empirical evidence on how grouping and categorising people remotely may affect them. Together they illuminate the ways in which big data is multifaceted and rich: by analysing location data that also has the dimension of time, we can analyse behaviour and action. Each case also involves research subjects who are unaware of the research and who are anonymous to the researcher, yet who may be significantly affected by interventions based on the data analysis. The cases described here deal with potential rather than actual harm, because the uses of data involved are still in development. The first refers to the identification of groups on the move through algorithmic profiling in the form of agent-based modelling; the second to identification as a group in a context of epidemiology, and the third to the identification of territory and its potential effects on those who live there. These cases are offered to make the point that while there are clear links between individual and group privacy and data protection issues, we have reached a stage in the development of data analytics where groups also need protection as entities, and this requires a new approach that goes beyond current approaches to data protection.

2.2 Background: The Current Uses of Big Data Analytics to Identify Groups in LMICs

People in LMICs have always been identified, categorised and sorted as groups through large-scale data, just like those in high-income countries. Traditional survey methods usually identify individuals as part of households, businesses or other conscious forms of grouping, using the group as a way to locate subjects and thus achieve legibility on the individual level. Such surveys are often conducted by states or public authorities, with the aim of identifying needs and distributing resources. In the case of LMICs they may also be conducted by international organisations or bilateral donors (e.g. UNICEF's Multiple Indicator Cluster Surveys, the InDepth Network's health and demographic surveillance system and USAID's Demographic and Health Surveys). Over recent decades, however, another mode of data gathering has become possible: identifying people indirectly through the data produced by various communications and sensor technologies. This data is becoming increasingly important as a way of gathering information on the characteristics of developing countries when conventional survey data is sparse or lacking (Blumenstock et al. 2014). Because most of this type of data is collected by corporations and is therefore proprietary, new institutions are evolving to provide access to and analyse it, such as the UN's Global Pulse initiative (Global Pulse 2013).

Although the new digital datasets may be a powerful source of information on LMIC populations, the implications of this new type of identifiability for people's legibility are huge and ethically charged, for reasons explored in the case studies below. 'Big data'[2] generated by citizens of LMICs is generally not subject to meaningful protections – for example, 8 out of 55 Sub-Saharan African countries had data protection legislation in place in 2013 (Greenleaf 2013) – and the data protection instruments that apply to multinational corporations gathering data in the EU or US have no traction regarding data gathered elsewhere in the world (Taylor 2016a). Those who work with these data sources from LMICs, however, rely on anonymisation and aggregation as ways to deflect harm from individuals (Global Pulse 2014). For instance, when mobile network provider Orange shared five million subscribers' calling records from Côte d'Ivoire in 2013 (Blondel et al. 2012) those records were both anonymised and blurred, so that the researchers who received the dataset had no way to make out individual subscribers' identities. Yet Sharad and Danezis (2013: 2) show how, in this dataset, even an anonymous individual who happens to produce high call traffic can lead to the spatial tracking of the social grouping he or she belongs to, using local information such as traffic patterns and the addresses of businesses (ibid.).

Data analytics can also tell us the characteristics of anonymous groups of people, either by inference based on the characteristics of a surveyed group within the larger dataset (Blumenstock 2012), or by observed network structure. Caughlin et al. (2013: 1) note that homophily, the principle that people are likely to interact with others who are similar to them, means that from people's communication networks we can identify their contacts' likely 'ethnicity, gender, income, political views and more'. In the case of the data used by the UN Global Pulse initiative, its director noted that:

> Even if you are looking at purely anonymized data on the use of mobile phones, carriers could predict your age to within in some cases plus or minus one year with over 70 percent accuracy. They can predict your gender with between 70 and 80 percent accuracy. One

[2] The focus here is on data that are remotely gathered and can therefore either be classed as *observed*, i.e. a byproduct of people's use of technology, or *inferred*, i.e. merged or linked from existing data sources through big data analytics (Hildebrandt 2013).

> carrier in Indonesia told us they can tell what your religion is by how you use your phone. You can see the population moving around. (Robert Kirkpatrick UN Global Pulse, 2012[3]).

Working with potentially sensitive datasets such as these is usually justified on the basis that the people in question can benefit directly from the analysis. This justification is double-edged, however, since the same data analytics that identify groups in order to protect them – for example, from disease transmission – may also be used to capture groups for particular purposes, such as to serve an adversary's political interests. One example of this is a data breach that occurred in Kenya during the 2012 election campaign where financial transfer data from the M-Pesa platform was accessed by adversaries and used to create false support for the registration of new political parties. In this case, people found they had contributed to the legitimacy of new political groupings without their knowledge (TechMtaa 2012) – something with enormous implications in a country which had been subject to electoral violence on a massive scale in its previous election, and where people were targeted based on their (perceived) political as well as tribal affiliation.

Nor is keeping data locked within the companies that generate them any guarantee against misuse. In a now notorious example, a psychological experiment was conducted using Facebook's platform during 2014 (Kramer et al. 2014) which showed that the proprietors of big data can influence people's mood on a mass scale. The researchers demonstrated that they could depress or elevate the mood of a massive group of subjects (in this case, two groups of 155,000) simultaneously by manipulating their news feeds on the social network, noting that doing so had the potential to affect public health and an unknown number of offline behaviours. It is important to note that the anonymisation of users in this case – even the researchers themselves had no way to identify their research subjects (International Business Times 2014) – did nothing to protect them from unethical research practices.

Cases of direct harm occurring on a group basis are not hard to find when one looks at areas of limited statehood or rule of law, which are often also lower-income countries. Groups, not individuals, were targeted in the election-related violence in Kenya in 2007–2008, in the Rwandan genocide of 1994 and in the conflict in the Central African Republic in 2013–2014. Similarly, political persecution may just as easily focus on groups as individuals, where a group can be identified as being oriented in a particular way. The sending of threatening SMS messages to mobile phone users engaged in political demonstrations, whether through network hacking as in Ukraine in late 2013 or by constraining network providers to send messages to their subscribers as in Egypt in 2011, was aimed at spreading fear on a group level, rather than identifying individuals for suppression. In fact, in many cases it is precisely being identified as part of a group which may make individuals most vulnerable, since a broad sweep is harder to avoid than individual targeting.

The ethical difficulty with this type of analysis is that it is a powerful tool for good or harm depending on the analyst. An adversary may use it to locate and wipe

[3] Robert Kirkpatrick, interview with *Global Observatory*, 5/11/2012. Accessed online 19/2/2015 at http://theglobalobservatory.org/interviews/377-robert-kirkpatrick-director-of-un-global-pulse-on-the-value-of-big-data.html

out a group, or alternatively it could be used to identify groups for protection. An example of the former would be in situations of ethnic or political violence, where it is valuable to be able to identify a dissident group that is holding meetings in a particular place, or to target a religious or ethnic group regardless of the identity of the individuals that compose it. During the Rwandan genocide, for example, violence was based purely on perceived ethnic group membership and not on individual identity or behaviour. An example of protection includes the use of mobile phone calling data in Haiti after the 2010 earthquake, where a group of researchers tracked the migrants fleeing the capital city in order to target cholera prevention measures (Bengtsson et al. 2011). The latter case demonstrates the flexible nature of an algorithmic grouping: 'the group' was not a stable entity in terms of spatial location or social ties, but a temporary definition based solely on people's propensity to move away from a particular geographical point.

These very different misuses of data are mentioned here because although they centre on the illegitimate use of personal data, they illustrate a new order of problem that is separate from the exposure of personal identity. The political hackers in Kenya wanted to increase their parties' numbers by accessing and appropriating the 'data doubles' (Haggerty and Ericson 2000) of large quantities of people, not to reach them individually and persuade them to vote one way or another. M-Pesa's dataset was attractive because it presented just such large numbers which could be grouped at will by the adversary. The Facebook researchers similarly were interested in the group, not the individual: they note that the kind of hypothesis they address could not be tested empirically before the era of big data because such large groupings for experimental purposes were not possible. In each case, individual identity was irrelevant to the objectives of those manipulating the data – the researchers in the Facebook study justified their use of data with reference to Facebook's user agreement, which assures users that their data may be used internally for research purposes, i.e. not exposed publicly.

Existing privacy and data protection provisions such as the EU 1995 directive[4] and its successor, the General Data Protection Regulation[5] focus on the potential for harm through identification: 'the principles of protection must apply to any information concerning an identified or identifiable person' (preamble, paragraph 26). The methods used in big data analytics bypass this problem and instead create a new one, where people may be acted upon in potentially harmful ways without their identity being exposed at all. The principle of privacy is just one of those at work in legal instruments such as the 1995 directive: the instrument is also concerned with protecting rights and freedoms, several of which are breached when they are unwittingly grouped for political purposes or subjected to psychological experiments. However, the framing of privacy and data protection solely around the individual

[4] Directive, E. U. (1995). 95/46/EC of the European Parliament and of the Council of 24 October 1995 on the protection of individuals with regard to the processing of personal data and on the free movement of such data. Official Journal of the EC, 23(6).

[5] General Data Protection Regulation 5853/12.

inevitably distracts from, and may even give rise to, problems involving groups profiled anonymously from within huge digital datasets.

In the following sections, three cases are outlined in which group identity, defined by big data analytics, can become the identifiable characteristic of individuals and may determine their treatment by authorities.

2.3 Case 1. Groups in Motion: Big Data as Ground Truth

Barocas and Nissenbaum (2014) warn that 'even when individuals are not "identifiable", they may still be "reachable", … may still be subject to consequential inferences and predictions taken on that basis.' In various academic disciplines including geography and urban planning, research is evolving along just these lines toward using sources of big data that reflect people's ordinary activities as a form of ground truth – information against which the behaviour of models can be checked. As ground truth, this data then comes to underpin Agent Based Models (ABMs), which facilitate the mapping and prediction of behaviour such as human mobility – for example, particular groups' propensity to migrate, or their spatial trajectory when they do move.

Big data reflecting people's movements, in particular, is a powerful basis for informing agent-based models because it offers a complex and granular picture of what is occurring in real space. Mobile phone data in particular is useful as ground truth for modelling, because it can show individuals responding to events in real time on a mass scale. The new sources of big data allow both a more granular level of ground truth for models on the small (city) scale, and the possibility to extend this to a larger scale as well, since the kind of analytics that can be used to model flows of people through cities can also be extended to model flows of people between countries or regions.

An example of the way cities are using data to model and predict people's movements can be drawn from a project undertaken in one European capital city during 2014, which involved tracking flows of people from the periphery into the city centre. For this project, a combination of sensors was used to give three types of data point: first, cameras with facial recognition software were installed along the main routes into the centre; next, wifi counters were set up to collect the signals from electronic devices (mainly mobile phones); and last, mobile phone GPS data was collected (via an intermediary) from a major mobile network provider for the period of the project. This set of data sources provided a way to disambiguate the individuals moving through the area (i.e. to tell whether a signal that appeared and reappeared was emitted by one person or several), to see the volume and speed of human traffic over the course of the 2 months, and to track whether individuals were making the same trip once or multiple times. It also showed which shops they visited, where they paused or took public transport, and what kind of group they were travelling in (families, tours, individuals or other groupings).

This level of sensor data, brought together from multiple sources as in the case of this project, creates data doubles which – although at first glance anonymous – are composed of various characteristics which might lead to people being treated one way rather than another. Any tracking software used over a period of days creates a unique signature for an individual (de Montjoye et al. 2013), which is considered a privacy risk by urban authorities conducting projects such as the one described here. On the group level, however, this is not considered as sensitive because it does not make individuals identifiable. Despite this, combined with cameras and wifi data it was possible to know people's movements at a new depth of detail in ways which could give city authorities the ability to manipulate their behaviour beyond simply movement. If added into a model as the basis for understanding how different groups travel through the city, this detail made it possible to predict how people of different ages, origins and social configurations would move through urban space, what attracts their interest, what makes them take one route rather than another, and how they influence each other's movement and behaviour. The kinds of conclusions that can be drawn from the data, then, are valuable not only to city authorities wanting to predict which areas will become crowded at which times, but also to firms interested in gaining people's attention and law enforcement interested in who might be creating trouble, all on the group level.

Beyond this urban scenario, there are many types of big data from the developing world that could be fed into such models in order to predict behaviours more broadly. These include financial transactions conducted over mobile phones, movement details from GPS sensors in various types of device, utility usage such as water or electricity in a smart-metered system (smart meters are being used on the district level, if not for individual houses, even in slums to enable water authorities to understand where water is being siphoned off illegally), internet search trends and social media postings. The work of Global Pulse, a UN initiative, shows how authorities are becoming interested in creating predictive models for entire countries that can then show what will happen when there is a particular type of shock, or when a shock occurs in a particular place. ABMs are also powerful because they can be used to explore counterfactuals: what would happen if an event happened in one place, as opposed to another, or at one time instead of another? This can be done particularly clearly with big data because they provide an unprecedented detail and granularity with regard to people's activities.

The key difference between big-data-based models and what preceded them is knowledge discovery – the practice of finding completely unforeseen questions and issues through data mining, rather than using data to answer known questions or test known hypotheses. The whole of big data, for researchers, behaves considerably differently from the sum of its parts. As data mining becomes increasingly integrated into modelling techniques, models will predict people's behaviour and movements in greater detail and with multiple scenarios because the researcher can alter the parameters and rerun the model in different ways to bring up different possible behaviours. In terms of planning for emergencies and population movements, large datasets combining different sources of data will determine the kind of preparation we make, and the kind of built environment in which we live. However, these decisions have their own politics. The built environment and the planned city are designed to make certain types of movement possible and visible, and to discourage other types. We may want to expose the movements of some groups, make some more easy to police or surveil, and thus to control. They can also be used on a smaller scale – company ID tags that track a worker's movements through the building, or RFID technology which tracks the movements of the objects we use, can reflect movements in a way that makes it possible to police the group.

This kind of research becomes even more of a risky proposition for groups when we consider it on the international scale. The years since 2009 have seen a great increase in the amount of research that aims to track people's movement in the developing world. Because the main technology offering tracking possibilities in low- and middle-income countries is the mobile phone, this research focuses on mobile phone traces. de Montjoye et al. (2013) have shown that mobile data can be an extraordinarily efficient way of identifying human mobility in the context of both ad hoc groupings and social networks, and identifying when these groups move simultaneously. Taking this a step further, the 2013 D4D challenge (Orange 2015) served to demonstrate how mobile data can already predict mobility, and what a useful tool it could become for either planning or preventing human movements. Examples of ABMs focusing on migration dynamics already exist (e.g. Kniveton et al. 2011), and the trajectory of big data research shows an evolution from one-time data analysis such as tracking epidemics (Bengtsson et al. 2011) to the broader use of mobile traces as parameters for agent-based models (ABMs) which may be used to predict mobility (Frias-Martinez et al. 2011; Pindolia et al. 2012).

The risk attached to such practices is not the uncovering of individual identity. Digital traces from the phones migrants carry with them may be traceable to registered SIM cards in their countries of origin, but in fact names and addresses at the place of origin would not be important in comparison to the ability to track the movement of the group. Unwanted migrants may be caught on an individual basis, but are resisted by receiving states on the group level. For example, if a group of people carrying mobile phones are attempting to cross the Mediterranean and enter the EU, they can be tracked in real time by anyone with access to the data. The data will also show their place of origin via the phone's record of their original network provider, and (if it is a smart phone) will show the networks they have connected to along the way, making it possible to identify whether they have taken an overland route and are therefore likely to be undocumented. The phone data thus conveys how many people are moving, where they come from, and the route they have taken. The value of this information is its potential to identify where a group is on its way, and to understand whether this is likely to be a group which might be able to claim asylum and which would include minors and highly vulnerable people (for example Syrians fleeing violence), or whether it is likely to be a group of economic migrants (for example from West Africa).

The potential ethical problem with this type of modelling is that it can provide new sources of information for a pressing policy interest: how to 'manage' the migration of the poor towards richer countries. In a realist interpretation, such data might allow a receiving state to determine whom to rescue and whom to ignore, or might lead to the choice to let groups of migrants drift into other states' territorial waters in order to avoid responsibility. In either scenario, the individual identity of the migrants is immaterial to the state's decision, while the identification of the group would be the basis for its survival or abandonment. In a world where the governments of higher-income countries have a strong interest in controlling mobility, and specifically in predicting, tracking and preventing unauthorised migration flows towards their borders, this potential for identifying the group becomes invaluable.

The added ability to predict mobility adds to the potential ethical problems with this scenario: what if certain conditions are met in a 'sending' country that make it likely, according to a model, that a population will be squeezed out of their territory and towards another country? How should authorities in each country respond to predicted, rather than realised, movement? If mobile phone data increasingly crosses its current institutional boundaries – as is likely, since 'function creep with data technologies can be taken as a given'[6] – and is used either as real-time surveillance data or in agent-based models it clearly has the potential to help governments preempt undocumented migration. This potential makes it hard to imagine that this immense capability to visualise and track migration will not cross the boundary from care to control.

2.4 Case 2. Groups and Disease Transmission: Big Data as Tracking Technology

Data scientists and epidemiologists have collaborated to use newly available sources of digital data in order to track and predict outbreaks of disease. Perhaps the best-publicised example of disease tracking is Google Flu Trends (Dugas et al. 2012), which uses Google search records to track flu infections over the winter. The accuracy of the Google Flu Trends model is checked by comparison with doctors' reports and hospitals' admissions data – checks which revealed that the Flu Trends algorithm overestimated flu cases during the 2013–2014 flu season at double their true level, probably because it counted searches by people seeking to distinguish cold from flu symptoms (Lazer et al. 2014). More recently new data sources where this kind of ground truth is not available – or at least not until the epidemic has passed – have been used in higher-stakes scenarios to track a variety of life-threatening diseases in developing countries, notably cholera in the wake of the 2010 Haiti earthquake (Bengtsson et al. 2011) and the use of network analysis to track malaria transmission (Tatem et al. 2009). This kind of tracking via data is particularly useful in the developing world, where public health infrastructure and capacity are often lacking and where the new data analytics can provide an unprecedented real-time element to authorities looking to control outbreaks of diseases.

[6] Dennis Broeders, keynote presentation, *Responsible Data for Humanitarian Response* conference, February 24–25, Leiden University, held at Foreign Ministry of the Netherlands.

For epidemiologists, tracking disease is a step towards predicting its spread, and ground truth is essential in doing so. The 2013 Data for Development Challenge (Orange 2015) produced 14 research papers focused on modelling the spread of epidemics in a West African context, incorporating information tools to control the movement of the disease via identifiable social groups. These focused on malaria and HIV (e.g. Brdar et al. 2015), but were not informed by actual disease information reflecting the period in question. In contrast, Wesolowski et al's study in Kenya (Wesolowski et al. 2012) combined mobile phone data with existing longitudinal survey data on malaria prevalence to identify the particular types of mobility, and mobile groups, indicated as principally responsible for parasite importation between regions. Wesolowski et al. note that their analysis could lead either to local efforts to control malaria through measures such as drugs and bednets, or could contribute to larger, population-based strategies on the part of authorities:

> Control-program activities targeting the large volumes of human traffic between regions that we have identified here will be completely different from those that concentrate on local transmission hot spots, focusing on communicating risks to travelers to alter their behaviors, restricting travel patterns, and/or conducting routine surveillance in high-risk areas. (Wesolowski et al. 2012)

These projected results of the researchers' analysis suggest that mobile phone data is becoming recognised as a possible motivator for restriction of movement and for surveillance. Given how much of Sub-Saharan Africa is a high-risk area for malaria (WHO 2014), if this possibility were actually realised then travel restrictions would be in place across much of the continent. Equally the option of surveilling those areas would impose an impossible burden on public health authorities, to the point where a meaningful effort to wipe out the disease would require participation from other authorities such as the military or law enforcement – with attendant risks of other activities being picked up apart from malaria transmission.

These concerns become sharpened by the availability of big data. Models such as Wesolowski et al.'s are verified by using historical information after the event – i.e. once the actual path of disease transmission has been tracked. This allows researchers to demonstrate that their model provides the best fit with what actually happened. With the new sources of big data, however, this can occur in real time. By using mobile phone GPS data, social media data or other forms of information that update as the epidemic progresses, it becomes possible to continually orient and re-orient the model so that it can adapt to predict the movement of people, and thus disease, with the maximum possible accuracy. Thus big data can become a new form of ground truth, and one which allows the researcher to work at a distance rather than seeking confirmation either from local fieldwork or from survey-based methods that involve individuals engaging with the model's 'agents' on the local level.

The threat to the group, rather than the individual, arises from the processes of quarantine that become possible once such data is available. It also arises from the type of data available about different populations. Mobile phone data from the West African locations of the 2014–2015 Ebola outbreak is of a different quality to mobile phone traces from a high-income location such as London or New York. If an outbreak occurred in such a location, it would be possible (at least in theory) given current technology to track transmission of the disease on the individual level. With continually updating GPS details from mobile phones and the ability to analyse individuals' social networks and communication dynamics, it would be possible for public health authorities to see movement and contact on a granular level, and to track and quarantine people individually if necessary. In contrast, mobile phone data from Liberia is less granular – though far more accurate than other currently available data on population movements such as satellite images. Because smartphones are not yet common in Sub-Saharan Africa (Telecoms.com 2014), GPS data would not be available and datasets would instead reflect which antenna a phone was closest to at a given time. In urban areas there are more antennas, providing greater specificity, while in rural areas there are far fewer, meaning that data on people's location gets fuzzier the further they move from a city.

This lack of granularity would be replicated in any decisions to quarantine people based on such data. In the 2014–2015 outbreak potential Ebola sufferers were quarantined based on location in a decision-making process that has changed little in a thousand years – a slum area would be fenced off and guards placed at the gates to keep inhabitants in (NY Times 2014). This approach necessarily has a high error rate in terms of identifying people with the disease, and involves people catching the disease who otherwise would not due to being contained in proximity with sufferers. Basic mobile phone data would not necessarily solve this problem due to the lack of granular detail available: instead it might increase authorities' perception of the risk of infection without narrowing down who might have contracted the disease, giving support to decisions to contain the sick and the healthy together by force.

A reliance on big data analytics, then, has the potential to remove an epidemic such as Ebola from its political and societal context (a lack of resources on the part of health authorities, and a lack of incentive to act on the part of unaffected countries) and place it instead in the data domain, which has solvable problems (a lack of data can be solved by gaining access to more and better data). In this domain, the political and human problems of quarantine decisions instead become a data problem: whom to confine and where. In this case, if the wrong decisions are taken based on biased or unclear data, the newest technology could only facilitate medieval decision-making processes targeted at groups rather than individuals.

2.5 Case 3. Drone Data and the Cross-Contextual Flow of Information

Large numbers of people in the world live in areas that are poorly mapped. Regions with low economic activity and few international connections, in particular, have historically not provided a strong incentive for cartographers. This dearth of spatial information is also reinforced by a lack of usership for such information – if one's home region has not been well base-mapped, then digital navigation tools, crowd-mapping and other technologies will not be able to layer on top to add value and depth of information. Mapping is therefore subject to a limited Matthew effect (one with no effect on the poor) where the better-mapped generate more input in terms of spatial information, and the poorly-mapped remain information-poor regardless of the development of new technologies. Exceptions are projects such as the Humanitarian Open Street Map project, which collaborates with development institutions such as the World Bank to map areas where better spatial information would aid development interventions.

On a higher institutional level, however, things may be changing. Powerful imaging technologies such as satellites and drones are increasingly being focused on LMICs for commercial, development and humanitarian purposes. Non-military drones are advocated as a way of gaining access to rural and remote customers (BBC News 2012); are deployed by the UN in peacekeeping operations in the Democratic Republic of Congo (Crowe 2013) and by entrepreneurs for humanitarian response after natural disasters (Churchill 2014). The World Bank is seeking to demonstrate the potential of drones for predictive and planning purposes 'in many sectors including: cadastral mapping/registration, infrastructure projects (roads, energy and dams), urban planning, and disaster risk management' (Volkmann 2014). Each of these LMIC spatial data projects has a different population focus (consumers, rebel militias, fleeing crisis victims, and farmers to name a few), a different stated purpose, and is made up of different institutional configurations, and is subject to different forms of governance.

Where these new sources of knowledge coincide with old sources of conflict, however, new ethical questions arise. A case in point is Harvard's Signal Program on Human Security and Technology, part of the Harvard Humanitarian Initiative. The Signal Program operates a project named Mass Atrocity Remote Sensing: analysing satellite and other spatial data to identify forensic evidence of alleged massacres. Their work on alleged atrocities in Sudan since the separatist conflict of 2011 demonstrates several fundamental problems that arise when researchers gain access to unprecedentedly detailed and granular data on territories in conflict. The research was retrospective, but findings were updated daily and could provide an ongoing picture of what was occurring in Sudan. The first problem encountered by the project was that it appeared to be providing intelligence to those conducting the atrocities. Program Director Nathaniel Raymond noted that

> we saw circumstantial and anecdotal evidence that people were making decisions on the ground, for good or for bad, based on our reporting of the satellite imagery analysis.[7]

Program researchers found that unknown adversaries appeared to be hacking into their communications, both on an individual level via team members' phones and on an institutional level, accessing their database through their servers.

[7] Interview with Nathaniel Raymond, Director, Signal Program on Human Security and Technology, Harvard University (25.2.2015).

Adversaries also appeared to target directly local people who were communicating with the researchers, identifying them through their use of portable satellite broadband connectors (BGANs). This combination of factors allowed hostile actors on the ground to use the research project's data and communications as a way of targeting their enemy more effectively.

A parallel problem encountered by the project was the lack of an ethical framework to deal with the conflicts arising from their data analysis. Raymond found that data protection frameworks focusing on individual identifiability became irrelevant in the context of large-scale satellite imagery processing:

> It's about demographic threat now. Before we thought of privacy and consent in terms of individuated risk and responsibilities to individuals. You look at the ICRC [International Committee of the Red Cross] professional standards for protection work – they are great on paper, [but] they're anachronistic. They're anachronistic because they are seeking to prevent harm through individual data release or [where] a group of individuals have their data released. They're not focused on the attack model. The attack model is identification of demographic group.[8]

This example demonstrates that once the data reflects the group and not the individual, adversaries may seek to silence people on the group level. Under these circumstances the threat can no longer be mitigated by ethical standards developed to protect individuals reporting abuses. Furthermore, what Raymond refers to as the 'tempo' of human rights reporting is changing with the new sources of data: instead of a report based on individual sources being compiled and published some time after an event, constantly updating population-level data makes a daily reporting schedule possible, and increases the chance that those committing atrocities will seek to complete their actions quickly, to get ahead of any reporting cycle which might lead to accountability.

> in fact we don't know how we may have sped [the adversary's] decision-making because they think, hey, we're on candid camera, we had better get in and out from robbing the convenience store as quickly as possible. (Nathaniel Raymond)[9]

The Harvard project is an extreme example of how new data sources may make populations vulnerable through making them visible, but also an example of how different technologies of visibility make people differently vulnerable. The most precise satellite data available to civilian researchers at present is at a resolution of 50 cm (Raymond et al. 2014). At this level,

> Crowds of livestock and people can sometimes be visible, though the exact composition, size, and object type of these crowd configurations cannot be reliably determined. Additionally, "micro interactions", such as the movement of small groups of individuals and the positioning of small weapons, cannot be reliably identified and tracked. (Raymond et al. 2014 [p.40])

At such a resolution, tracking and understanding actions on the ground involves visualising groups rather than individuals – and as the Harvard project demonstrates,

[8] Raymond interview, (25.2.2015).

[9] Raymond interview, (25.2.2015).

may provoke a response from hostile parties on the group level rather than the individual.

The problem of remote sensing groups and their territory is not restricted to human rights reporting, but has the potential to create new forms of surveillance of populations who were previously effectively invisible on the international level. Drones and satellites pick up all activities, not only those the user is targeting, so that projects conceived with beneficial motives may still provide records that can be used for other purposes. While watching militias form and move in African conflict zones, the UN can also see the activities of the entire population, and while seeking particular consumers in remote areas, commercial drones will inevitably pick up other activities, locations and movements. These datasets, like other big data about human activities, will almost inevitably be subject to function creep. They will make it possible to identify, sort, categorise and predict with relation to populations who often have very limited access to their own governments, or which may want to stay anonymous if they are endangered by being recognised. In some cases anonymity is necessary for a group's wellbeing as a strategy to preserve land, culture and autonomy.

2.6 Discussion

The cases outlined above demonstrate how it is necessary to reconceptualise the risk of data harms to include the problem of the group, not only the individual. This is especially true in environments such as LMICs where privacy and data protection rules are often not yet clearly set out or enforceable, and where states (who often, under current law, have the responsibility to pursue cases of data misuse) may themselves be the perpetrators of harm. As well as the risk to established groups such as separatists, however, new risks may be posed that operate across established categories. Big data analytics specifically offer the potential to discover new information, identify patterns and predict behaviour, and thus to algorithmically delineate entirely new groups which may be cross-ethnic and cross-border.

Researchers who specialise in LMIC issues, and especially social scientists focusing on developing countries, may be the least well-placed to pick up on these issues. Development Studies research in particular tends to address groups as defined by borders and social identities. International and non-governmental organisations focusing on LMIC populations in a human rights or aid context may also fail to recognise the new categories of risk due to the prevalence of firm, but outdated, ethical guidelines. The beneficial aims of these various types of work also conspire to make it less likely that researchers in these fields will seek to understand how new forms of data may create new risks to research subjects. The terminology of 'development', 'aid' and 'humanitarian response' insulates researchers from criticism and accountability, and thus disincentivises them from seeking out the problems inherent in the research practices evolving around big data.

The sources of big data are also difficult for researchers to manage with regard to consent and awareness on the part of research subjects. Consent on the group level has not been addressed either in the technology industry or in academic research, and therefore researchers engaging in practices as varied as crowdsourcing information, performing satellite data analytics or processing mobile phone traces must re-think the entire way that consent to the use of data is conceptualised and given:

> How do you assess buy-in when fundamentally, because you can't do individual consent, you are talking about community consent? And if you are doing community consent, that is gendered, class-based and ethnic in a way that presents even more dimensions of problems. (Nathaniel Raymond)[10]

A further problem is how consent should operate in the context of fast-moving events and real-time data. A network operator may donate data as an emergency response at a time of crisis, as happened with the Haiti earthquake and cholera epidemic (Bengtsson et al. 2011), and that data may be dealt with responsibly, as it was by the research team in question. However, as the power of such large-scale data becomes more widely understood, there are calls for making data available by default to international researchers, not only local authorities in the case of emergencies (Economist 2014).

The fastest-moving events currently visualised using big data are arguably those involving epidemics. These present the possibility that the new sources of data may enable a particularly extreme response depending on the perceived seriousness of the epidemic. The scenario of widespread travel restrictions and surveillance, for example, is relatively unlikely in the case of malaria, which is both survivable (although highly dangerous, especially for children) and treatable. A different calculus of risk applies in the case of Ebola, which has a fatality rate of up to 80 % (Team 2014) and where there is, so far, no reliable mode of treatment. During the 2014–2015 Ebola epidemic in West Africa public health authorities were stretched far beyond capacity, resources were lacking on the international scale, and there were calls for the release of mobile phone datasets (e.g. Talbot 2013) as a way to help authorities overcome this challenge.

However, the extreme fear and urgency surrounding the Ebola crisis and the predominance of international research teams in the debate over data availability give rise to questions regarding the way that data might lead to the targeting of disease through groups, rather than addressing individuals as patients (or potential patients). For instance, research produced by the first Orange Data for Development challenge demonstrated that international researchers often conceived of LMIC environments as similar to HIC ones in terms of the way authorities and populations would react to an epidemic. For example, their models (e.g. Lima et al. 2013) tended to assume a population of informed individuals, signed up to digital information networks in order to receive real-time information from authorities. The researchers envisage (2013: 1) that 'a collaborative effort leveraging individual social ties can

[10] Nathaniel Raymond interview (25.2.2015).

be effective in propagating effective information (i.e. a sort of "immunizing information") to a widespread audience.'

In contrast, the Ebola epidemic presented a scenario where sufferers were perceived as a group, and where due to local conditions of limited technological access and education, that groupness lent itself to rumour and misinformation, collective fear, and consequently to coercion and violence on the part of authorities, including forcible quarantine where the lives of the uninfected were endangered (NY Times 2014). In this type of scenario, big data models built to facilitate individuals' well-being and autonomy instead would constitute perfect tools for mass control and surveillance. The kind of model made possible by mobile phone data, for example, allows authorities to identify networked groups (such as commuters on a transport line, or those who work in a particular area) as potential carriers of a disease, and therefore raises the incentive to control and confine them regardless of their actual infectious status. The utility of mobile phone data in the case of this epidemic has since been critically assessed (McDonald 2016) with the finding that Ebola's particular characteristics mean that tracking people through mobile data on a group level would be ineffective in combating the spread of the disease, and that individually identifiable mobile traces would in fact be the only remotely collected data that could help chart its spread.

The logic of this call for mobile data releases is that international researchers have greater capacity than local ones, and will therefore provide more insights from the same dataset. As the examples offered above imply, however, the greatest problems may arise precisely because of the release of data to international researchers rather than local ones, for several reasons. First, because those researchers are inevitably operating without first-hand experience of the territory, the crisis in question or the people affected. They may therefore misunderstand the risks inherent in a particular dataset or analytical practice. Second, because data has an almost infinite half-life. Regardless of ethical research frameworks that aim to stop the reuse and sharing of data beyond specific users, it is in the nature of data to replicate, and of technological infrastructures to facilitate its replication. Digital data is increasingly difficult to delete entirely. Once stored, copied or transmitted it exists in multiple locations which often extend across international borders and may form grey areas in terms of data governance (as in the case of the cloud computing which currently facilitates much of big-data analytics). In the absence of appropriate ethical frameworks to deal with the problem of exposing groups through data analytics, then, the data will continue to spread and multiply, becoming ever-more linkable, mergeable and creating new forms of risk as researchers become desensitised to its conditions of origin.

2.7 Conclusion

This chapter has outlined ethical problems with the definition of groups through data technologies in three areas: modelling and predicting mobility through agent-based models; predicting disease transmission through network analyses, and visualisation technologies that provide information on previously hard-to-research populations. It has identified several new problems arising with regard to the use of new data technologies to map, track, sort and analyse people on a group level. First, that of data analytics in the area of 'knowledge discovery': data mining or other techniques that create algorithmic groupings and that seek to predict the movements or activities of groups defined this way. One example of this would be a model that identifies who is likely to move and in which direction in the case of a particular climate event. This may involve big data as a way of parameterising a predictive model, or as a way of informing it in real time. As ubiquitous computing becomes more of a reality, sources of data for this kind of modelling will become increasingly available at lower cost and with fewer restrictions. The risks posed to groups by this kind of algorithmic analytics are particularly clear on the political level: if unwelcome movements can be predicted, authorities can step in before people become defined as refugees, asylum-seekers or other problematic categories that award the right to move.

A further risk of this kind of predictive modelling is that it tends to blur categories, 'seeing' people according to their propensity to behave in a certain way, rather than as individuals. At a resolution of 50 centimetres no one has an individual identity, and where data mining techniques are concerned, the kind of personal information that can currently be protected by law is wholly irrelevant.

This chapter has looked at contexts where consent for data use is usually not sought or possible: epidemics, crises, conflicts and remote sensing. The questions that arise, however, show that consent is highly relevant. Human rights, crisis response and development research efforts all have in common the aim of producing actionable information. If research, however remote, is aimed at impacting the condition of its subjects, but is conducted without attention to people's consent and awareness, it will raise ethical questions – as seen in Facebook's experiment, despite the company's claim that people had consented by using the service in the first place. If the 'group' is too unwieldy, too analytically unstable or too remote to consult and gain consent from, should this not constitute a major problem in terms of conducting the research? However, under current legal conditions the opposite is true: where subjects' names are not attached to their data, they are considered anonymous and the use of their information to be innocuous. In fact, as this chapter has shown, the reverse may be true.

The potential risks and harms outlined in this analysis all relate to the consequences of drawing conclusions about a given group based on assumptions drawn from other groups. As such, they are problems with treating the group as a category – a definition that flattens out difference – rather than as a spectrum of types which will include outliers to whom the intervention or categorisation will not apply. We see these types of categorisations in real life every day: some are merely inconvenient, such as badly targeted direct advertising. Other generalisations may be fatal, for instance if one is a civilian caught in an airstrike targeted at a military area. Most are on a spectrum between these extremes, as with the examples of migration and quarantine offered above, and most raise issues of both privacy and data protection because they incorporate problems both of visibility and identification, and of protection from intervention. These problems point to the need for a new ethical approach to research with regard to group-level information. Demographic-level research is fundamentally changing and evolving to offer ever more possibilities for categorisation, by a wider group of potential analysts. As the sources and types of data change, so too do the conditions of their use. No longer can the authorities demand that a researcher who wants population-level information undergo a vetting process and be shut into a room with a census database, forbidding them to take the data home with them. Instead of being shared vertically in an institutional hierarchy where the data owner has power over the researcher, the new forms of digital data are shared horizontally. They are crowdsourced and crowd-analysed, shared, reused, replicating into the cloud and onto individual hard disks, under the label of humanitarian response, development hacking and poverty mapping.

In none of these circumstances is consent seen as possible, nor has it been conceptualised on the group level where the most serious risks seem to lie. This chapter suggests that these two problems may be related. If we can conceptualise how data analytics affect groups, we may find it impossible to proceed without some kind of ethical dialogue with those groups. At the same time, without an imperative for consent to researchers' use of big data, the notion of the group is allowed to remain both conceptually fuzzy and practically challenging. The institutions currently working with and advocating the use of digital traces from LIMCs tend to emphasise the importance of traditional conceptions of privacy, focusing on personal information and the debate around pseudonymisation and other forms of identity-blurring (e.g. Global Pulse 2014; GSMA 2011). In contrast, the Harvard Humanitarian Initiative points clearly to new problems that are hard to classify as relating to individual privacy (Raymond et al. 2014).

Thus, to answer one of the questions posed at the start of the chapter, it seems that the problem of group profiling contains recognisable elements of both privacy and data protection problems: people's fundamental right to autonomy is being affected, but they are also consequently being made vulnerable to discrimination and personal danger. Given that the problems outlined above are an inherent issue with big data analytics in general, however, privacy may provide the best conceptual 'hook' for understanding and addressing these problems. The right to privacy has arguably always been used to get to thorny and hard-to-define problems because it touches on various more concrete rights – those of autonomy and the right to intellectual freedom, freedom from surveillance and interference, and the right to behave in ways that may be inconvenient for the authorities. With regard to LMICs, these freedoms are central to resisting the kind of threat potentially facilitated by new visualisation and data analytical possibilities.

The second overarching question of this chapter was whether, by resolving the problem of individual identifiability, we resolve that of groups. The cases presented above suggest that we do not – the group issue makes it necessary to look beyond individual identification to larger issues of the rights that are abrogated when data is misused. This leads to a larger question of accountability for data misuse, which is currently not occurring. When research operates in the international sphere essentially free from accountability to local populations, as is common practice in the development and humanitarian spheres, an ethical framework must address the problem of operating, and seeing, remotely. If the group is becoming the only category available, standards of ethical behaviour must be reworked and evolved to match current reality.

If this happens, however, it will address an established and ongoing challenge: how to make those who remotely visualise and affect populations accountable to those populations. Thus addressing the problem of group privacy may enhance individual privacy and other rights in two important ways: first, by making it necessary to find ways to contact research subjects and find out whether they consent – or dealing with the fact that they cannot be located and cannot therefore consent. And second, as a result of that process, by connecting the researcher to the reality of their research subjects and thus necessitating a broader, more risk-averse approach that focuses on the contextual understanding of risks. This has much in common with the contextual approach to privacy advocated by Nissenbaum (2010) – but with the caveat that data about LMICs always seems to be subject to exceptionalist claims based on need and crisis, and that this is unlikely to change until global power asymmetries do (Taylor 2016b).

Finally, the last overarching question: if we solve the problem of group privacy in one place, does that lead to a more universal solution? The examples provided here suggest that this depends on the place. If we solve this problem for places with the least geopolitical power we go a long way towards solving it for places with less extreme risks. The measures which will prevent authorities in one place from targeting groups for violence will also prevent those in another from targeting them for discriminatory health insurance premiums. The reverse is not true, however. If data is not addressed as a source of power and as a right in itself, then people in disempowered places cannot hope to figure in the decisions of data controllers.

Bibliography

Barocas, S., and H. Nissenbaum. 2014. Big data's end run around anonymity and consent. In *Privacy, big data, and the public good: Frameworks for engagement*, ed. J. Lane, V. Stodden, S. Bender, and H. Nissenbaum. Cambridge: Cambridge University Press.

BBC News. 2012. *Matternet: Swapping roads for flying drones*. Accessed 15 Aug 2014 at http://www.bbc.com/future/story/20120209-i-say-to-you-today-i-hover-dream.

Bengtsson, L., X. Lu, A. Thorson, R. Garfield, and J. von Schreeb. 2011. Improved response to disasters and outbreaks by tracking population movements with mobile phone network data: A post-earthquake geospatial study in Haiti. *PLoS Medicine* 88: e1001083.

Bettencourt, L.M. 2014. The uses of big data in cities. *Big Data* 2(1): 12–22.

Blondel, V.D., M. Esch, C. Chan, F. Clérot, P. Deville, E. Huens, … C. Ziemlicki. 2012. Data for development: The d4d challenge on mobile phone data. arXiv preprint arXiv:1210.0137.

Bloustein, E.J. 1978. *Individual and group privacy*. New Brunswick: Transaction Publishers.

Blumenstock, J.E. 2012. Inferring patterns of internal migration from mobile phone call records: Evidence from Rwanda. *Information Technology for Development* 18(2): 107–125.

Blumenstock, J.E., R. Chokkalingam, V. Gaikwad, and S. Kondepudi. 2014. Probabilistic inference of unknown locations: Exploiting collective behavior when individual data is scarce. In *Proceedings of the fifth ACM symposium on computing for development*, December, 103–112. ACM.

Brdar, S., K. Gavric, D. Culibrk, and V. Crnojevic. 2015. Unveiling spatial epidemiology of HIV with mobile phone data. arXiv preprint arXiv:1503.06575.

Caughlin, T.T., N. Ruktanonchai, M.A. Acevedo, K.K. Lopiano, O. Prosper, et al. 2013. Place-based attributes predict community membership in a mobile phone communication network. *PLoS ONE* 8(2): e56057.

Churchill, E. 2014. Philippines: Drones and spreadsheets to the rescue. http://blogs.worldbank.org/eastasiapacific/philippines-drones-and-spreadsheets-rescue.

Crowe, A. 2013. *United Nations' drones: A sign of what's to come?* Accessed 15 Aug 2014 at https://www.privacyinternational.org/blog/united-nations-drones-a-sign-of-whats-to-come.

de Montjoye, Y.A., C.A. Hidalgo, M. Verleysen, and V.D. Blondel. 2013. Unique in the crowd: The privacy bounds of human mobility. *Scientific Reports* 3: 1376.

Dugas, A.F., Y.H. Hsieh, S.R. Levin, J.M. Pines, D.P. Mareiniss, A. Mohareb, … R.E. Rothman. 2012. Google flu trends: Correlation with emergency department influenza rates and crowding metrics. *Clinical Infectious Diseases* 54(4): 463–469.

Economist. 2014. *Waiting on hold.* October 25, 2014. Accessed online 16 Mar 2015 at http://www.economist.com/news/science-and-technology/21627557-mobile-phone-records-would-help-combat-ebola-epidemic-getting-look.

Eindhoven News. 2014 *Orange scent should stop aggression on Stratumseind.* Acccessed online 6 Oct 2015 at http://www.eindhovennews.nl/news/26208-orange-scent-should-stop-aggression-on-stratumseind.html.

Frias-Martinez, E., G. Williamson, and V. Frias-Martinez. 2011, October. An agent-based model of epidemic spread using human mobility and social network information. In *Privacy, security, risk and trust (passat), 2011 ieee third international conference on and 2011 ieee third international conference on social computing (socialcom)*, 57–64. IEEE.

Global Pulse. 2013. *United Nations Global Pulse Information Sheet.* Accessed online 19 Nov 2015 at: www.unglobalpulse.org.

Global Pulse. 2014. *Frequently asked questions: Data privacy.* Online brief. Accessed 19 Nov 2015 at: http://www.unglobalpulse.org/about/faqs.

Greenleaf, G. 2013. Sheherezade and the 101 data privacy laws: Origins, significance and global trajectories. *Journal of Law, Information & Science* 23(1): 4–49.

GSMA. 2011. *Mobile privacy principles.* Accessed online 19 Nov 15 at: http://www.gsma.com/publicpolicy/wp-content/uploads/2012/03/gsmaprivacyprinciples2012.pdf.

Haggerty, K.D., and R.V. Ericson. 2000. The surveillant assemblage. *The British Journal of Sociology* 51(4): 605–622.

Hildebrandt, M. 2013. Slaves to big data. Or are we?. Idp. *Revista De Internet, Derecho y Política* 16.

Hildebrandt, M., and S. Gutwirth (eds.). 2008. *Profiling the European citizen.* Heidelberg: Springer.

International Business Times. 2014. *Were you a subject in facebook's mood experiment? You'll never know.* Accessed 19 Nov 2015 at: http://www.ibtimes.com/were-you-subject-facebooks-mood-experiment-youll-never-know-1616970.

ITU. 2013. *The world in 2013. International Telecommunications Union.* Accessed 19 Nov 2015 at http://www.itu.int/en/ITU-D/Statistics/Documents/facts/ICTFactsFigures2013.pdf.

Jaquet-Chiffelle, D.O. 2007. Direct and indirect profiling in the light of virtual persons. In: *Profiling the European Citizen*, eds. M. Hildebrandt and S. Gutwirth, 34–43. Springer.

Jerven, M. 2013. *Poor numbers: How we are misled by African development statistics and what to do about it.* Ithaca: Cornell University Press.

Kniveton, D., C. Smith, and S. Wood. 2011. Agent-based model simulations of future changes in migration flows for Burkina Faso. *Global Environmental Change* 21: S34–S40.

Kramer, A.D.I., J.E. Guillory, and J.T. Hancock. 2014. Experimental evidence of massive-scale emotional contagion through social networks. *Proceedings of the National Academy of Sciences* 111(24): 8788–8790. Available at: http://www.pnas.org/cgi/doi/10.1073/pnas.1320040111.

Lazer, D., R. Kennedy, G. King, and A. Vespignani. 2014. The parable of google flu: Traps in big data analysis. *Science* 343(6176): 1203–1205.

Lima, A., M. De Domenico, V. Pejovic, and M. Musolesi. 2013. Exploiting cellular data for disease containment and information campaigns strategies in country-wide epidemics. *arXiv preprint arXiv:1306.4534.*

Lyon, D. 2014. Surveillance, Snowden, and big data: Capacities, consequences, critique. *Big Data & Society*, 1(2), 2053951714541861.

Mao, H., X. Shuai, Y.Y. Ahn, and J. Bollen. 2013. Mobile communications reveal the regional economy in Côte d' Ivoire. In *Proceedings of the 3rd conference on the analysis of mobile phone datasets* (NetMob).

McDonald, Sean M. 2016. *Ebola: A big data disaster. Privacy, property, and the law of disaster experimentation*. CIS Papers 2016.01, March 2016. http://cis-india.org/papers/ebola-a-big-data-disaster.

Nissenbaum, H. 2010. *Privacy in context: Technology, policy, and the integrity of social life*. Palo Alto: Stanford University Press.

NY Times. 2014. *Clashes erupt as liberia sets an ebola quarantine*. Accessed online 17 Mar 2015 at http://www.nytimes.com/2014/08/21/world/africa/ebola-outbreak-liberia-quarantine.html?_r=0.

Orange. 2015. *Data for development*. http://www.d4d.orange.com/en/Accueil.

Pindolia, D.K., A.J. Garcia, A. Wesolowski, D.L. Smith, C.O. Buckee, A.M. Noor, … A.J. Tatem. 2012. Human movement data for malaria control and elimination strategic planning. *Malaria Journal*, 11, 205.

Raymond, N.A., B.L. Card, and I.L. Baker. 2014. A new forensics: Developing standard remote sensing methodologies to detect and document mass atrocities. *Genocide Studies and Prevention: An International Journal* 8(3): 33–48.

Richards, N.M. 2013. The dangers of surveillance. *Harvard Law Review* 126: 1934.

Sharad, K., and G. Danezis. 2013 De-anonymizing D4D datasets. In *Proceedings of the 13th privacy enhancing technologies symposium*. July 10–12, 2013, Bloomington, Indiana, USA.

Strandburg, K. 2014. Monitoring, datafication and consent: Legal approaches to privacy in the big data context. In *Privacy, big data, and the public good: Frameworks for engagement*, ed. J. Lane, V. Stodden, S. Bender, and H. Nissenbaum. Cambridge: Cambridge University Press.

Talbot, D. 2013. Cell-phone data might help predict ebola's spread. *MIT Technology Review*. Accessed 17 Sept 2014 at http://www.technologyreview.com/news/530296/cell-phone-data-might-help-predict-ebolas-spread/.

Tatem, A.J., Y. Qiu, D.L. Smith, O. Sabot, A.S. Ali, and B. Moonen. 2009. The use of mobile phone data for the estimation of the travel patterns and imported Plasmodium falciparum rates among Zanzibar residents. *Malaria Journal*, 8, 287.

Taylor, L. 2016a. Data subjects or data citizens? Addressing the global regulatory challenge of big data. In *Freedom and property of information: The philosophy of law meets the philosophy of technology*, ed. M. Hildebrandt and B. van den Berg. New York: Routledge.

Taylor, L. 2016b. No place to hide? The ethics and analytics of tracking mobility using mobile phone data. *Environment and Planning D: Society and Space* 34(2): 319–336.

Team, W.E.R. 2014. Ebola virus disease in West Africa—The first 9 months of the epidemic and forward projections. *New England Journal of Medicine* 371(16): 1481–1495.

TechMtaa. 2012. *Outcry over claims that M-Pesa details are used 'To Register' party members*. Accessed 28 July 2014 at http://www.techmtaa.com/2012/01/16/outcry-over-claims-that-m-pesa-details-are-used-to-register-party-members/.

Telecoms.com. 2014. *Africa gets smart: Continent prepares for device revolution*. Accessed online 17 Mar 2015 at http://telecoms.com/opinion/africa-gets-smart-continent-prepares-for-device-revolution/.

Volkmann, W. 2014. *Small drones for progress in international development projects*. Post on LinkedIn, April 24, 2014. Accessed 15 Aug 2014 at https://www.linkedin.com/today/post/article/20140424005308-38421723-small-drones-for-progress-in-international-development-projects.

Wesolowski, A., N. Eagle, A.J. Tatem, D.L. Smith, A.M. Noor, R.W. Snow, and C.O. Buckee. 2012. Quantifying the impact of human mobility on malaria. *Science* 338(6104): 267–270.

Wesolowski, A., C.O. Buckee, L. Bengtsson, E. Wetter, X. Lu, and A.J. Tatem. 2014. Commentary: Containing the Ebola outbreak–the potential and challenge of mobile network data. *PLOS Current Outbreaks.*

WHO. 2014. *World malaria report 2013*. Geneva: World Health Organization.

World Economic Forum. 2014. *Data-driven development: Pathways for progress*. New York: World Economic Forum.

Chapter 3
Group Privacy in the Age of Big Data

Lanah Kammourieh, Thomas Baar, Jos Berens, Emmanuel Letouzé, Julia Manske, John Palmer, David Sangokoya, and Patrick Vinck

With contributions from
Augustin Chaintreau, Yves-Alexandre de Montjoye, Natalie Shoup, Paula Kift

Abstract Until now, privacy protections have focused on guaranteeing individuals a measure of control over information relating to themselves. However, in the digital age, this protection has become less effective since data is constantly collected and stored in ways that make it difficult for the individual to have control over each piece of information. Furthermore, the information communicated by an individual, when processed in conjunction with other data points, may allow potentially harmful inferences to be drawn about other individuals and the groups to which they may belong. The potential of Big Data to harm groups, particularly in fragile contexts or areas of weak statehood, therefore raises a number of questions which this chapter seeks to explore: is there such a thing as group privacy, distinct from individual privacy? Is group privacy a workable concept? If so, should it be a legally enforceable right and how can it be protected? We begin by exploring various concepts of privacy and group; then discuss how to affirm and protect group privacy through a combination of traditional levers of power and better data management, security, and literacy.

Keywords Privacy • Group privacy • Human rights • Group rights • Discrimination • Bias • Algorithmic transparency

---

L. Kammourieh (✉) • T. Baar • J. Berens • J. Manske • J. Palmer • D. Sangokoya • P. Vinck
Data-Pop Alliance, New York, USA
e-mail: lkammourieh@datapopalliance.org

E. Letouzé
Data-Pop Alliance, New York, USA

The Data-Pop Alliance / MIT Media Lab, New York / Boston, USA

L. Taylor et al. (eds.), *Group Privacy*, Philosophical Studies Series 126,
DOI 10.1007/978-3-319-46608-8_3

3.1 What Is a Group?

What do we talk about when we talk about groups? First, the term "group" in its ordinary meaning designates "a number of persons or things," that is to say a class or unit made up of more than one person or thing.[1] While the separation between a group and its individual members might seem obvious, we will see that it is no longer as neat as it might have been in the past. With Big Data analysis, an individual's habits and characteristics can increasingly be taken to represent a class of similar individuals and, on their own, suffice to draw conclusions about a group.

Secondly, when we refer to groups of people, we usually think of a social, religious, or ethnic group, or a structured organization such as a company, association, or political party – in short, we imagine people who have formed explicit ties, whether legal or otherwise, that bind them together. Technology changes this, too. With Big Data analysis, individuals' data can be aggregated with unprecedented ease. Once individual information has been turned into a data set, subsets may easily be extracted from this – thus grouping together, based on certain common traits or practices, individuals who have no awareness of being bound by these similarities.

Before defining group privacy, we must therefore look more closely at what a group has usually meant and how legal systems have envisaged groups up to this day (A). We will then address the changes created by Big Data and the question of whether there is still a clear distinction between an individual and a group (B).

3.1.1 Traditional Notions of Groups: Self-Proclaimed and Self-Aware

The traditional notion of a social group involves some degree of shared perception of the group as a group being defined by its members, outsiders, or both. In other words, the traditional social group exists as part of a collective consciousness. The members of a self-aware, or "active" social group identify themselves as such and proclaim the group's existence. Their identities are often shaped by the perception and treatment of the group by the rest of society. At the same time, such external social perceptions may also form the basis of what we may call a "passive" social

[1] We use the notion of group in the ordinary meaning of the term. It must be noted that related notions exist in specific disciplines. Logic, linguistics, and computer programming all refer to the type-token relationship to distinguish between a class or concept (the type) and the objects that instantiate it (the tokens). Similarly, mathematics refers to the set-element relationship; in this framework, it is possible for a set to have only one element, in which case the set is called a singleton. We use the notions of group and members, or group and individuals, in order to connote the cross-disciplinary nature and human focus of our inquiry, and to enable us to formulate recommendations with a policy-making and legal reach. In the ordinary meaning of these terms, a group is usually made up of more than one member, and we will focus on these situations.

group: one treated as a group by society without its members actually identifying themselves as such. For example, populations considered vulnerable or fragile have been denoted as such based on the absence of relative social dimensions (such as income, agency, and resilience), while the members themselves do not necessarily perceive themselves as part of the group.

The characteristics underlying shared perceptions of groups include socially constructed categories like race and ethnicity, as well as physical, psychological, or behavioral characteristics such as sex, political opinion, or union participation. Many self-aware groups, of course, are constituted intentionally, and take the form of organizations or communities, in which the bonds and relationships between members may be the most important characteristics. Such intentional groups often have legal personality in domestic and international legal contexts. These groups may even themselves be sources of law or regulation, as in the case of states and sub-state entities like provinces, cities, or, in some customary law systems, tribes.

Aside from possessing legal personality and, sometimes, law-making functions, traditional groups and group-related concepts have played other important roles in a number of areas of the law, whether in the international or domestic legal framework. For example, fundamental norms of equality and the prohibition of discrimination focus on the risks and harms of individuals being treated differently based on their membership in certain groups.[2] Another example is refugee law, which protects those who are persecuted on account of group membership.[3] These are all cases of the law protecting *individuals* from group-related harms; but there are also many ways in which the law protects and gives rights to groups *as groups*.

For example, the International Covenant on Civil and Political Rights (ICCPR) protects the family as "the natural and fundamental group unit of society."[4] The Genocide Convention requires states to prevent and punish certain conduct aimed at the destruction of a national, ethnic, racial, or religious group "as such."[5] The right to self-determination is held by "peoples," generally defined in ethnic, linguistic, or religious terms.[6] Minority rights, although mostly expressed in terms of individual group members, have some elements that can be exercised only at the group level.[7] Similarly, trade unions are directly accorded specific rights in international and domestic law to organize and function freely.[8]

More recent emerging norms of international law arguably address group rights as well. The U.N. General Assembly's Basic Principles on the Right to a Remedy

---

[2] *See, e.g.* International Covenant on Civil and Political Rights, Art. 26.

[3] Convention Relating to the Status of Refugees, Arts. 1A, 33.

[4] Art. 23.1.

[5] Art. II.

[6] E.g., ICCPR, Art. 1.

[7] For instance, article 27 of the ICCPR prohibits states with ethnic, religious or linguistic minorities from denying members of these minorities the right, "*in community with the other members of their group*, to enjoy their own culture, to profess and practise their own religion, or to use their own language" (emphasis added).

[8] E.g. International Covenant on Economic, Social and Cultural Rights, Art. 8.1(b)–(c).

and Reparations suggest that the right to reparations may be held by groups of people who have been targeted collectively.[9] Similarly, the International Criminal Court's Rules of Procedure and Evidence suggest that this court interprets its statute so as to allow for collective reparations. Such collective reparations have already been awarded in a number of decisions by the Inter-American Court of Human Rights.[10]

Box 1: History of Group Rights
At few moments in history did the obligation to protect members of groups become as pressing as in the aftermath of World War II. The Convention on the Prevention and Punishment of the Crime of Genocide (CPPCG), adopted by the UN General Assembly in the course of the Holocaust, explicitly refers to discrimination and violation against national, ethnical, religious or racial groups. It was during that same time, triggered by the same concern, that the right to privacy was acknowledged as a fundamental human right – stated in Article 12 of the Universal Declaration of Human Rights. It was regarded as a central pillar of democratic societies particularly because it reinforces other rights, such as freedom of expression and information, as well as freedom of association. As a result, it has been embedded in international human rights law and domestic laws as well as policies in democratic societies and beyond.

Insofar as the international and domestic legal systems have taken groups into account, they have done so for traditional self-aware, or "active," groups. However, this approach is interesting in that it could potentially extend to groups that are not necessarily able to represent themselves (whether as a matter of obtaining legal capacity or even as a simple matter of internal organization and cohesion). This could encompass traditionally passive groups as well as the new types of passive groups created by Big Data.

3.1.2 Big Data: New Grounds for Identifying Groups

A group is constituted by a number of individuals classed together. As seen above, the classification of a number of individuals as a human group has traditionally occurred through a social construct setting its members apart from other individuals and/or groups of individuals. The group's existence could be enforced internally, i.e. by the members of the group itself, and/or externally, i.e. by outsiders to the group.

[9] Principles 8, 22(g).

[10] See generally Friedrich Rosenfeld, "Collective reparation for victims of armed conflict," 92 International Review of the Red Cross 731 (2010).

Subsequently, the classification might be acknowledged or disputed: being a member of a group, or being excluded from it, can have significant implications for individuals.[11]

Groups have always been formed by classification based on commonalities perceived by members and outsiders. But with the advent of a digitized society, groups are now being defined in ways different than before – no longer by mere human perception, but, for example, with the use of algorithms. As a result, it is important to rethink the definition of what a group is, and to understand the new ways in which we find commonalities. More specifically, it is key to assess the impact of new epistemic practices related to data analysis on group identification.

The increased availability of personal data results in a wealth of data points on human beings. Rich data sets, in turn, can be used to infer commonalities between individuals. As the traces we leave behind become virtually innumerable, the common characteristics based on which we can identify groups are multiplied. Big Data does not refer only to the overwhelming wealth of digital data now available, but also to the development of new tools and methodologies to process this data. Through machine learning, it is now possible to infer information and draw knowledge from vast amounts of unstructured data. Pattern recognition facilitates the discovery of previously imperceptible interrelations within datasets and, as such, creates new means for identifying and grouping individuals. As data and information retrieval processes become increasingly sophisticated, so does the process of group identification. Groups can now seem to automatically present themselves within data, even as the picture of the individual members remains fuzzy. Big Data thus changes what a group is and, in the same sweep, what an individual is.

The application of automated forms of data analytics, such as machine learning and data mining, can affect the ways in which we identify and think of groups in four main ways:

- First, data analytics can help us find out new things about pre-existing, self-defined "active" groups. Although the group might have been formed and defined before any data was collected, we now have the capacity to infer new information from data about these groups without having any pre-defined hypothesis in place.
- Secondly, we might come to identify previously non-apparent groups on the basis of certain pre-defined parameters. For example, a data analyst may choose one characteristic – such as pattern of telephone usage – and query his database to find seemingly unconnected users who exhibit similar behaviors.
- Thirdly, without defining any parameters or characteristics upfront, we might discover groups through new analytical approaches. This can lead to the identification of new groups on the basis of sets of characteristics previously unknown even to the data analyst.

[11] *See, e.g., Bowker and Star's work on the devastating impact of classifying humans by race under South Africa's Apartheid regime.* Geoffrey C. Bowker and Susan Star, *Sorting Things Out*, Cambridge, MA, The MIT Press, 1999.

- Fourth and lastly, while using such analytical processes, there will be an increasing risk of algorithms identifying new groups as a step in the analytic process, even as data scientists remain unaware of it. The claims resulting from the analysis might in turn affect or harm these groups, even as the group itself remains latent – with neither the group members identifying themselves as such nor the data scientist "seeing" that the group has been extracted from the data. This is possible in two case scenarios: either the group has been identified within the data mining process itself but has not become apparent to the analyst; or a group classification has been enforced through the analytical process by the choice of certain data which is non-representative or biased in some way.

Box 2: Understanding the Complexity of Groups in the Age of Big Data: Black Twitter

"Black Twitter" refers to a group of internet users active on Twitter as a platform for global group discussion, advocacy, and biting commentary on Black experience both in the U.S. and around the world. The group has been noted for its impact in the U.S. racial discourse, particularly with the emergence of the #BlackLivesMatter campaign. Viral tweets such as #BringBackOurGirls, #AliveWhileBlack and #OscarsSoWhite have also been described as part of the group's influential portfolio of trending hashtags.[12]

This has resulted in attempts by marketing agencies and news organizations to analyze who is involved in these conversations and what is the impact of the group's activities. State and federal government agencies have also attempted to infer and track the activity of users linked to the organization of protests and rallies for the Black Lives Matter movement.[13]

Yet Black Twitter as a group has no defining labels or clear indicators, making it difficult for outsiders to infer involvement of any one user. Indeed, the mere use of a hashtag as an indicator of group involvement is problematic, due to the hashtag's ephemerality and lack of clarity; one-time use of a hashtag can denote mere group affinity just as it can denote self-identification.

Attempts to identify "Black Twitter" have often resulted in crude inferences involving multiple users and tweets that may not have been involved in the network at all. Depending on the nature and intent of the analyst seeking to establish the classification, this can have a range of harms for those thus misidentified or not self-identifying, including targeted solicitation of unwanted services, latent discrimination as a by-product of associative algorithmic decision-making or biased data collection, and government surveillance of civil rights activities.

[12] http://harmony-institute.org/latest/2013/08/06/blacktwitter-a-networked-cultural-identity/

[13] http://www.oregonlive.com/politics/index.ssf/2015/11/black_lives_matter_oregon_just.html

Big Data thus provides new approaches with which groups can be formed. Where group classification seemed to hinge on the salience of certain commonalities between individuals, Big Data makes the grounds upon which we can identify new groups increasingly imperceptible – first to the group's members themselves, who might be classed together without ever knowing they share common characteristics; and then potentially to data analysts as well. The use of more sophisticated and complex technologies makes the nature of the connections between different data points, and the impact of these connections on group identification processes, increasingly opaque. In this context of increased automation of knowledge, an epistemic shift might occur in which the analyst's consciousness of information extraction will be blurred. Groups might no longer be classified based on the perception of certain observers, but through seemingly obscured algorithmic processes. This incomplete awareness of how and on which grounds group identification takes place could lead to an epistemic dependence on processes we might no longer fully understand.

Understanding this recent evolution is key in order to evaluate the new privacy risks created by Big Data, and to examine the notion of group privacy in particular.

3.2 What Is Group Privacy?

The concept of privacy is notoriously difficult to define and has varied and sometimes conflicting interpretations. We choose to view privacy as a facet of human dignity: one's right to have a measure of knowledge and control over what information is made public about oneself (A). Applying this concept to groups is a complex operation (B). First, it must be determined when individual privacy ceases to offer sufficient protection to members of a group; when the group might be at risk even as its members' individual privacy is protected; so that it becomes apparent that something called group privacy, separate and different from individual privacy, is at stake and requires protection. Second, we must examine the practicalities of creating and protecting a privacy right in an international or domestic legal framework. Indeed, in order to hold rights, groups must have legal personality. We must ask what rights, if any, can be given to "passive" groups who are not self-aware and organized but merely extracted from the data. In configurations where a group privacy right cannot be granted, can other forms of protection be found to prevent abuses?

3.2.1 Challenging Traditional Notions and Protections of Privacy

3.2.1.1 The Shifting Ontology of Privacy

The translation, implementation and observance of the right to privacy in the 'digital age' have received much attention in the academic, corporate and public sectors over recent years. The appointment of Professor Cannataci as U.N. Special

Rapporteur on the Right to Privacy is only one among many indicators of the amount of attention that this right currently receives.[14] A number of elements of this discussion build on fundamental notions of privacy, and a review of these elements is essential to understanding the current evolution of the debate.

Privacy remains operationally a "fuzzy concept"[15]: there is no broad consensus on what exactly privacy is, and consequently on what a right to privacy should protect. Daniel Solove has underlined "the great difficulty in reaching a satisfying definition of privacy," a discontent that "persists even though the concern over privacy has escalated into an essential issue for freedom and democracy."[16] In the United States, the right to individual privacy emerged as protection against state infringement on personal life, as well as in reaction to the emergence of photography and a more enterprising, and sometimes intrusive, press. It has famously been conceived of as "the right to be let alone."[17] But privacy has also been conceived in myriad other ways. Solove thus lists various other conceptions drawn from a wide array of academic works in disciplines including law, philosophy, psychology, and sociology. Conceptions of privacy include the ability to shield oneself from the unwanted access by others; the right to keep secrets, that is to conceal certain things from others; the ability to exercise control over information about oneself; the protection of one's personhood, individuality, and dignity; and control over the intimate aspects of one's life.[18]

In another seminal article, James Whitman tackles the "disconcertingly diverse forms" of privacy – identifying two broad "cultures" of privacy, one leaning more towards liberty (from the state) and the other towards dignity (the right to one's own image and reputation).[19] Both aspects are, of course, indispensable to a healthy democracy. As Harry Lewis argues with regard to anonymity,[20] the ability to operate outside of the scrutiny and judgment of the public is essential to developing a counter-narrative on major societal issues. Throughout history, individuals and groups have needed to retain spheres of privacy as protection against the surveillance powers of the state. Uprisings such as the American Revolution and more recent Arab Spring movements would not have been possible had their developments been fully known by the established political regimes. The right of citizens to a private sphere is, in part, what allows for counter-narratives to be thought up and potentially lived out. Privacy has been a safeguard against state knowledge becom-

---

[14] *See* UN Resolution A/HRC/RES/28/16, to be found here: http://ap.ohchr.org/documents/dpage_e.aspx?si=A/HRC/RES/28/16.

[15] Solove 2008.

[16] Daniel Solove, "Conceptualizing Privacy", 90 California Law Review 1087, 2002, at 1089–1089.

[17] Samuel D. Warren and Louis D. Brandeis, « The Right to Privacy », *Harvard Law Review*, Volume IV, No 5, December 1890.. *See also* Solove 2008; vom Lehn 2014.

[18] Daniel Solove, "Conceptualizing Privacy", 90 California Law Review 1087, 2002, at 1094.

[19] James Q. Whitman, "Two Western Cultures of Privacy: Dignity Versus Liberty," 113 Yale Law Journal 1153, 2004. *See also* Bloustein 1964.

[20] 'Anonymity and Reason', *Privacy in the Modern Age* [ed. Rotenberg, Horwitz and Scott, The New Press, 2015].

ing too complete, and with it power becoming too absolute, making privacy one of the fundamentals of modern democracy for individuals and groups.

But privacy does not only affirm itself against the state and its surveillance powers. It can also protect people against the prying eyes of fellow citizens, as well as against corporations. Exactly how privacy deploys its protections varies from one legal system to another. Louis Brandeis and Samuel Warren's foundational article on privacy was written in reaction to a changing American press "overstepping in every direction the obvious bounds of propriety."[21] Paul Whitman has also pointed out, for example, that European privacy protections "are all rights to control your public image – rights to guarantee that people see you the way you want to be seen. They are, as it were, rights to be shielded against unwanted public exposure – to be spared public embarrassment or humiliation. The prime enemy [...] is the media." By contrast, he argues that the American right to privacy, "[a]t its conceptual core, [...] still takes much the form that it took in the eighteenth century: it is the right to freedom from intrusions by the state, especially in one's own home." But Whitman is quick to point out that these are, of course, relative and not absolute differences: both European and American legal systems incorporate privacy protections against both the state and the media – they simply do so in different proportions, and the emphasis each society chooses to place reveals much about its conception of privacy.

Lastly, the very need for a privacy right has also been the source of debate: some have argued against the existence of a separate right to privacy altogether,[22] whereas many others have shown continued support for the current inclusion of the right to privacy as a human right under international law (in the ICCPR, the Universal Declaration of Human Rights, and other instruments). The need for such legislation is particularly prominent in conflict-affected and terrorism-affected countries where security and government actors are increasingly deploying the use of new surveillance and data collection technologies by security and government actors has outpaced the development of guaranteed legal protections for data privacy.[23] While several developing countries such as Nigeria have constitutional provisions describing the privacy of citizens as "protected," these provisions lack specific details on the nature of these protections from the state, corporations or other citizens.[24]

While acknowledging the difficulty of reaching a definitive, *a priori* conception of privacy, we choose to focus on the approach of privacy as a form of dignity.[25] This conception of privacy aligns with Westin's definition of privacy as "*the claim of individuals, groups, or institutions to determine for themselves when, how, and to*

[21] Samuel D. Warren and Louis D. Brandeis, « The Right to Privacy », *Harvard Law Review*, Volume IV, No 5, December 1890.

[22] Yael Onn et al. 2012.

[23] See *Privacy International, "Lebanon: It's Time to You're your International Position on Privacy Into Action at the National Level," 2016.https://www.privacyinternational.org/node/586*

[24] Akinsuyi, F. Franklin. "Data Protection and Privacy Laws Nigeria, a Trillion Dollar Opportunity!!" Social Science Research Network. April 24, 2015. http://ssrn.com/abstract=2598603

[25] Solove 2008; Smith, Dinev and Xu 2011; Mayer-Schönberger & Cukier 2013.

what extent information about them is communicated to others".[26] It is particularly relevant to the discussion of the privacy risks posed by Big Data, where the transmission, collection, and analysis of information are key – and where the stakes extend far beyond freedom from the sole gaze of the state.

3.2.1.2 Data, Information, and Knowledge

Up until now, privacy protections, as diverse as their forms have been in international and various domestic legal frameworks, have focused on guaranteeing the individual a measure of control over information relating to him- or herself. In the digital age, this protection has become less effective. First, because data is constantly collected and stored outside the grasp of the individual, with the sheer multiplicity of digital "traces" left behind by each one of us making it more difficult to exercise control over each piece of information. Second, because this "raw" information provided by the data subject is no longer, in and of itself, the crux of the problem: such information communicated by an individual may well be harmless; but once processed, a great deal of valuable information can be inferred from it.

The works of Yves-Alexandre de Montjoye et al. have shown that it has become increasingly difficult, if not impossible, to anonymize a dataset (that is, to erase the names of the data subjects and ensure that these names cannot be found again by cross-referencing against other databases). This is due to three sea changes: the number of datasets that can be cross-referenced has grown; the data itself has become richer; and, as a result, the algorithms that succeeded in creating "noise" in datasets to prevent re-identification are no longer effective.[27]

In addition, and crucially, the very issue of anonymity and identification has become secondary: indeed, the richness of today's datasets mean they no longer allow us just to retrace an individual's name; data analysis can also allow us to make inferences about a data subject's personality, for example by detecting signs of extraversion or of neurosis. This is the meaning of metadata: the information revealed goes beyond that which is directly contained in the data.[28]

This shifts the locus of the problem. In the age of Big Data and information inferred *ab extra*, the traditional right to informational privacy no longer provides sufficient protection to the individual; it focuses solely on information collection rather than analysis, and can thus no longer be a fully effective instrument of control.

---

[26] Westin 1968.

[27] See also Paul Ohm, "Broken Promises of Privacy: Responding to the Surprising Failure of Anonymization", UCLE Law Review, Vol 57, p. 1701, 2010.

[28] de Montjoye, Yves-Alexandre, Samuel S. Wang and Alex Pentland, "On the Trusted Use of Large-Scale Personal Data," IEEE Data Engineering Bulletin, 35–4 (2012).

Box 3: 2013 Release of New York City Taxi Trip Data: Revealing Muslim Taxi Drivers with High Religiosity

As a response to a Freedom of Information Act (FOIA) request, the New York City Taxi and Livery Commission released all NYC taxi trip data from 2013, including trip dates and times (including pickup and drop-off), location coordinates, number of passengers and other variables.[29] Although taxi license and medallion numbers had been anonymized, users were able to infer PII by linking the dataset with geo-located social media data and metadata (e.g. using images of celebrities and time-stamped tweets to map the trips of celebrities).[30]

In addition to the individual privacy dilemmas resulting from the release, data users were also able to infer with a degree of accuracy whether a taxi driver was a devout Muslim or not by linking the pauses in their trips – to park, wash, and pray as a part of the ritual – with adherence to their regularly timed prayer times over the span of a year.[31] The group privacy implications arise in what can be inferred or projected as a result of denoting this classification as well as the variables associated with this group of drivers (and other passive groups). End-of-shift neighborhoods and groups visiting specific neighborhood mosques, for example, incorrectly classified as "radical" Muslim populations may become vulnerable to heightened surveillance and discrimination as a result of group inferences from the trip data. Additionally, the fact that this data exists and can one day be published as a result of a FOIA request warrants concerns related to group privacy as well.

These changes are significant for groups and individuals alike, and for the same reasons. Inherent to inferring information from data is pattern identification. These patterns are based on finding a property shared by a part of the dataset, and seeing how this property correlates with some other property. Identifying a property possessed by particular individuals means to create a group. When a second property is added, the group generally becomes smaller, as a lower number of individuals will share both properties. Continuing to add properties will generally cause the group to decrease in size. Add enough properties and the constituency of the group will end up one.

[29] Whong, Chris "Foiling NYC'S Taxi Trip Data" Chriswhong.Com. (2016) http://chriswhong.com/open-data/foil_nyc_taxi/

[30] "Riding With the Stars: Passenger Privacy in the NYC Taxicab Dataset." Neustar Research, September 15, 2014. http://research.neustar.biz/2014/09/15/riding-with-the-stars-passenger-privacy-in-the-nyc-taxicab-dataset/

[31] Berlee, Anna. "Using NYC Taxi Data to Identify Muslim Taxi Drivers." The Interdisciplinary Internet Institute. January 21, 2015. http://www.theiii.org/index.php/997/using-nyc-taxi-data-to-identify-muslim-taxi-drivers/

3.2.1.3 Demographically Identifiable Information (DII)

The size of the group in itself is information. More valuable, however, are the correlations that members of the group show with certain behaviors, characteristics, or other relevant aspects. As the members of the group are selected based on a higher number of properties, these correlations will be more likely to produce accurate descriptions and predictions regarding both the group itself and its individual members. In Nathaniel Raymond's earlier chapter, he defines this kind of information as "demographically identifiable information," or DII, namely: "either individual and/or aggregated data points that allow inferences to be drawn, enabling the classification, identification, and/or tracking of both named and/or unnamed individuals, groups of individuals, and/or multiple groups of individuals according to ethnicity, economic class, religion, gender, age, health condition, location, occupation, and/or other demographically defining factors."[32]

In isolation or through linking, DII comprises all forms of data in which the identification, classification, and tracking of demographic groups; this includes "personal identifiable information (PII), online data, geographic and geospatial data, environmental data, survey data, census data." As Raymond mentions, ethical implications resulting from DII can arise across the data chain (in collection, compilation, analytics and use) and be problematic "simply whether the possibility exists that it can be even created."

It should be noted that in the release of DII, whether intentionally or unintentionally, not all group privacy risks are equal. In some countries, group privacy violations mainly result in unwanted targeted ads and other inconveniences in customer experience. While these violations can and should warrant attention, the consequences and effects of group privacy violations for vulnerable groups, particularly those in fragile contexts and/or areas of limited statehood, can be potentially life-threatening. In these environments where the state lacks the capability and accountability mechanisms necessary to protect against privacy violations (both physical and digital), identification and association with groups facing demographic-based discrimination can result in unchecked aggression against both actual and perceived group members.

In other words, just as existing privacy rights are poorly equipped to address the richness and invasiveness of the inferences that can now be drawn about individuals, they also fail to account for the richness of inferences that can be drawn about groups, with particularly grave consequences for and effects on vulnerable populations.

[32] Raymond, Nathaniel. "Beyond 'Do No Harm' and Individual Consent: Reckoning with the Emerging Ethical Challenges of Civil Society's Use of Data." (forthcoming)

Box 4: Group Privacy, CDRs and Public Health Response

In the last few years call details record (CDR) data have been piloted for tracking disease dissemination and human mobility in order to improve response to outbreaks and disasters:

- Digicel Haiti provided Harvard researchers with anonymized datasets on the position of 1.9 million SIMs in Haiti from 42 days before the 2010 earthquake to 158 days afterwards. Nearly 200,000 SIMs that were present in Haiti's capital Port-au-Prince when the earthquake struck, had left 19 days post-earthquake. Just under a third of Port-au-Prince's inhabitants were mobile phone subscribers at the time of the earthquake, so this movement of SIMs equates to the movement of 630,000 people.[33]
- Harvard researchers analyzed a year of CDR data from Safaricom in order to map human mobility and its contribution to the spread of malaria in Kenya. By linking the data with national infectious disease data, researchers estimated the likelihood that specific map routes contributed to the spread of the disease.[34]
- The Namibia National Vector-borne Diseases Control Programme (NVDCP) used mobile phone data from MTC Namibia, in combination with surveillance data and satellite imagery, in order to analyze movement patterns for over a million people and map areas of malaria prevalence and risk.[35]

While the success of these pilots has raised further calls to the use of CDRs for humanitarian and development response, dilemmas remain in the efficacy of anonymity methods and the likelihood that these methods can protect against re-identification and the release of PII. In addition to these dilemmas, CDR data raises critical group privacy concerns as demographic categories such as ethnicity[36] and socioeconomic status,[37] can also be inferred.

---

[33] Bengtsson, Linus, et al. "Improved response to disasters and outbreaks by tracking population movements with mobile phone network data: a post-earthquake geospatial study in Haiti." *PLoS Med* 8.8 (2011): e1001083.

[34] "Using Cell Phone Data to Curb the Spread of Malaria." Harvard T.H. Chan School of Public Health News. October 11, 2012. http://www.hsph.harvard.edu/news/press-releases/cell-phone-data-malaria/

[35] Tatem, Andrew J., et al. "Integrating rapid risk mapping and mobile phone call record data for strategic malaria elimination planning." *Malaria journal*13.1 (2014): 1–16.

[36] Blumenstock, Joshua, et al. "Neighborhood and Network Segregation: Ethnic Homophily in a Silently Separate Society." *Proc. NetMob* (2015).

[37] See Decuyper, Adeline, et al. "Estimating food consumption and poverty indices with mobile phone data." *arXiv preprint arXiv:1412.2595* (2014); Smith, Christopher, Afra Mashhadi, and Licia Capra. "Ubiquitous sensing for mapping poverty in developing countries." *Paper submitted to the Orange D4D Challenge* (2013); Mao, Huina, et al. "Mobile communications reveal the regional economy in Côte d'Ivoire." *Proc. of NetMob* (2013).

3.2.1.4 Privacy Protection in the Age of Big Data

This epistemic shift requires us to rethink what it means to talk about privacy protection. As the ways in which data is transformed into information change, focusing on the outcomes of this analysis process will help to better safeguard privacy rights amid ever-evolving data use practices.

In this view, it is important to identify the various stages of information processing in which the right to privacy can be protected. The very first stage, that of defining what constitutes personal data, is a significant one. For example, ahead of the "trilogue" discussion of the E.U.'s proposed General Data Protection Regulation (GDPR) that is set to bind all business processing European citizens' data worldwide, the Article 29 Working Party, comprised of representatives from all E.U. Data Protection Authorities, tackled the process from its very basis and proposed to strengthen privacy protections by expanding the definition of "personal data". E.U. texts currently define personal data as "any information relating to an identified or identifiable natural person;"[38] now, the consensus text reached for the GDPR, due to come into force in 2018, has expanded the definition to cover a wider range of data types allowing for identification, including online identifiers or factors specific to the physical, physiological, genetic, mental, economic, cultural, or social identity of the data subject.[39]

The right to privacy can be protected at several further steps in the process from data to information. The most stringent way to protect the right to privacy would be to prevent any collection of data-points about individuals. This is, of course, unrealistic (and also undesirable). Another level of protection, used extensively over the past decade, consists in anonymizing a dataset by removing personally identifiable information during the data processing stage. However, as seen above, the advent of Big Data implies that this is no longer as effective: anonymization can render re-identification more difficult, but not impossible. A third level of protection could be to place a range of restrictions on the cross-referencing of datasets, even anonymized datasets, that in combination with each other could reveal sensitive information. Anonymity is no longer central. As it becomes near-impossible, and maybe even irrelevant, we must rethink what we intend to protect when we speak of protecting privacy. It could no longer be to prevent the collection of information, or even to prevent identification, but rather to find the means to block access to sensitive data or to prevent the cross-referencing that could produce sensitive information.

[38] Directive 95/46/EC of the European Parliament and of the Council of 24 October 1995 on the protection of individuals with regard to the processing of personal data and on the free movement of such data, http://eur-lex.europa.eu/legal-content/en/ALL/?uri=CELEX:31995L0046

[39] http://www.emeeting.europarl.europa.eu/committees/agenda/201512/LIBE/LIBE%282015%291217_1/sitt-1739884

To explore this approach further, the "spherical" or contextual notion of privacy introduced by Van den Hoven[40] and Nissenbaum,[41] respectively, is particularly useful: in this view, goods are allocated differently across different spheres – information is arguably such a type of good.[42] This requires defining spheres (contexts and purposes) in order to then define the circumstances in which data can be accessed by certain actors. Using this "spherical" or contextual notion, we could envision a variable-geometry privacy protection, taking on different strengths depending on the data subject, the data processor, the type of data, the type of use and the context in which it is used.

At the same time, the difficulties of this approach are clear: as new types of data and new practices continue to develop, it becomes very difficult to determine when and how to set these levels of privacy protection. We often do not know the magnitude and extent of privacy risks until the harms we feared have materialized; it is difficult to protect privacy by limiting access to data based on predictions of how it might be used.

One thing is certain: in the current context, privacy protection must combine limitations both in access to data and its use to extract certain types of information. In most cases, regulating the use of data will constitute adequate protection. However, for certain types of particularly sensitive data, it might be necessary to regulate the collection of the data itself, in order to reduce risks.

In this section, we have underlined both the importance of privacy to a democratic society and the difficulty of reaching a single definition of privacy. For proof, different societies have adopted different views of privacy, reflected in the different shapes privacy rights have taken in their respective legal systems. We have opted for a broad definition of privacy as a facet of human dignity and as the right to control the information one makes knowable about oneself. Second, we have indicated how the changes wrought by Big Data challenge us to rethink the ways in which to protect privacy: anonymization is no longer possible or maybe even relevant; and what must be protected is no longer "raw" data such as names but rather metadata, the valuable information that can be inferred from datasets. Third, we have identified some useful elements in thinking about adapting privacy protections to the age of Big Data, with one key element being the notion that privacy protection must combine limitations to data collection and access, with a regulation of data processing.

[40] Van Den Hoven, Jeroen. Information technology, privacy and the protection of personal data. Cambridge University Press, 2008.

[41] Nissenbaum, Helen. "Privacy as contextual integrity." Wash. L. Rev. 79 (2004): 119.

[42] For the analogy of information as product, see also Posner, Richard A. "The economics of privacy." The American economic review 71, no. 2 (1981): 405–409.

3.2.2 *The Content and Protection of Group Privacy*

Having established that traditional notions of individual privacy are no longer sufficient to cover the more diverse harms enabled by Big Data, this section explores in greater depth the content that the concept of "group privacy" might have, and how it could work to complete existing protections.

3.2.2.1 Group Privacy Is Not Reducible to Individual Privacy

Why is the concept of "group privacy" necessary at all in the age of Big Data? Would it not suffice to adapt individual privacy to the current technological context and simply strengthen it? If members of a group are all individually protected from unwanted intrusions and targeting, then isn't the group itself protected?

The answer is no. We have seen that new data collection and analysis capacities render the concept of groups more relevant than before: first, by making more information discoverable about existing groups; second, by increasing the ability to "extract" groups from data even as its members are unaware of their imputed membership; and third, potentially, by allowing for imperceptible "grouping" processes to occur at the data analysis stage, unbeknownst to the analyst himself. In this context, it is possible for individual privacy to be effectively protected while leaving the group itself insufficiently protected.

Imagine a situation in which each individual has shared his or her data knowingly and agreed, at the time, to the type of processing to be carried out. Now imagine that the lawfully obtained, lawfully processed set of personal data allows the analyst to draw sophisticated inferences – on, say, likely reactions to a certain event, or likely population movements – predicting the behavior of a group of individual data subjects *as a group*. Such inferences would be based not on analyzing past individual behaviors in order to predict future individual behaviors, but rather on comparing and contrasting the behaviors of all members of a group, the group having been defined on the basis of one or more shared characteristics.

Box 5: Group Privacy and Forced Migration
Refugee movements provide a ready hypothetical illustration of such dangers. Since the outbreak of the conflict in 2011, millions of Syrians have been displaced, either internally or internationally, fleeing their homes in search of safety. Consider the possibility of a town under assault, with groups of residents beginning to flee. The population can be broken down by religious beliefs, known political leanings, law enforcement history, and neighborhood of residence. The government may have ready access to such information, as well as the surveillance capability to monitor population movements in real

(continued)

Box 5 (continued)
time. Such data might reveal that 5 % of the town population left on week one of the assault, that nearly all members of the group belonged to the same religious community, that a significant percentage of them had previously been noted for anti-regime leanings, and that two neighborhoods of the town are overrepresented in the group. The following week, as conditions worsen, they are followed by a further 10 % of town residents sharing similar characteristics. Such data could easily be used to project population movements on week three, and change the parameters of military action accordingly. It is the analysis of the group as a group that could then allow the analyst to predict the behavior of a third wave of displacement. It might not be possible to say exactly which individual members will decide to leave next. But the inferences drawn can still conceivably put the group, as group, at risk, in a way that cannot be covered by ensuring each member's control over his or her individual data. It is in this sense that we can talk of a group privacy interest.

In some ways, of course, individual privacy can reinforce group privacy. The clearest example is perhaps the E.U.'s protection of "special categories of data" in the current Data Protection Directive, carried over in the GDPR, and which grants increased protection to specific categories of highly sensitive data: "revealing racial or ethnic origin, political opinions, religious or philosophical beliefs, trade union membership, and [...] the processing of data concerning health or sex life."[43] While this is a protection granted to the individual, its effect is also to protect specific groups that are more vulnerable to targeting. But while individual and group privacy do overlap, they remain two distinct sets of interests.

It is one thing to identify a group privacy interest, and another to create a group privacy right. This is the question we move on to now: if group privacy is increasingly at stake in the age of Big Data, can it be protected simply by creating a group privacy right, ideally enforceable in both international and domestic legal frameworks, just like individual privacy? The answer is more complex. As we have seen, Big Data means analysts and algorithms can now "see" groups where none were perceptible before; and not all of these groups can practicably wield rights.

3.2.2.2 Group Privacy as Dignity: Concepts of Self-Determination and Data Sovereignty as Applicable to Groups

We now turn to the question of the content that a group privacy right could have. Having opted for a view of privacy as dignity with, as its corollary, the ability to decide what one keeps private or makes public about oneself, we can attempt to

[43] Directive 95/46/EC, *supra* note 31, article 8.

flesh out what this might mean for a group. Two existing legal concepts are related to group dignity and can therefore help us approach a definition of group privacy: self-determination, and sovereignty.

The notion of self-determination can be traced back to the American Declaration of Independence and French revolution,[44] and culminated in the 1960s and 1970s decolonization movement. A core principle of international law, it designates the legal rights of peoples to decide their own destiny. Almost by definition, self-determination, at least in "its classical incarnation,"[45] can only be seen at play in revolutionary moments when "the people" coalesces as an actor to overthrow one government or form of government and opt for another. In more recent years, however, the notion has evolved into more than a mere vehicle of decolonization. To this classical notion of "external self-determination," some have opposed an "internal self-determination" that affords groups continuous political and social rights, including by allowing minority groups within a state to enjoy protection and a measure of autonomy.[46]

This is rooted in Article 1 of both the International Covenant on Civil and Political Rights (ICCPR) and International Covenant on Economic, Social, and Cultural Rights (ICESCR). The existence of such a right to internal self-determination remains a subject of debate. But the notion is relevant to help conceive of group privacy: first, because it is a right that belongs to groups as groups, underlining the unique interests that groups may have and formulating a particular right to help defend them. Second, because it encompasses and translates a notion of group dignity, one also at play in the concept (encountered above) of "informational self-determination."

Equally helpful is another fundamental principle of international law: that of sovereignty. On the international legal plane, sovereignty means that a state is not bound by rules that it has not itself accepted, it is not submitted to any superior authority. When states take on a new obligation, they wield their sovereignty in a way that restricts their own liberty, submitting them to rules in the way they have chosen. This legal fiction is a key instrument of the formal equality between states. But the term sovereignty has also been used in a different way: it was applied to peoples in the same decolonization period that saw the right to self-determination flourish. In 1962, a U.N. General Assembly resolution declared "the right of peoples and nations to permanent sovereignty over their natural wealth and resources."[47]

Both uses of the term "sovereignty" can help inform our discussion. They can help us imagine, by means of analogy, a privacy right that consists of retaining control over one's personal data unless one has explicitly consented to relinquishing it.

---

[44] Antonio Cassese, *Self-Determination of Peoples : A Legal Reappraisal*, Cambridge, Cambridge University Press, 1995, p. 11.

[45] Jonathan I. Charney, *Self-Determination : Chechnya, Kosovo, and East Timor*, Vanderilt Journal of Transnational Law, volume 34, p. 455.

[46] *Id.*

[47] G.A. Resolution 1803, U.N. GAOR, 17th session, Supp. No. 17, at 15, U.N. Doc A/5217 (1962).

They also help us conceive of a group's right to control certain resources that are crucial to the collective interest of the group.

However, while this notion is helpful, it faces two important limitations. The first is a factual limitation, identified in the previous subsection: while raw data is indeed a precious resource, it is no longer the sole key to effectively protecting privacy; the aggregation and analysis processes are just as important to regulate in order to effectively protect groups and must therefore also be addressed. The second is a legal limitation: the rights of self-determination and sovereignty described above are wielded by a legal subject – the state, the people. Indeed, legal personality is the very capacity to hold rights: it has been defined as "the particular device by which the law creates or recognizes units to which it ascribes certain powers or capacities."[48] Exercising rights also, of course, requires self-awareness: the rights granted to peoples can only be asserted by a group that identifies itself as such, and state sovereignty can only be asserted by political leaders aware that their office allows them to commit and bind the state. The analogies to self-determination and sovereignty can thus help us conceive, at least partially, of the privacy right of self-aware, active groups – but they cannot be applied to the myriad "passive" groups extracted by the data analysis process.

How, then, can we envisage the protection of passive groups? In the impossibility of granting them sovereignty over or a right to control group data, their protection should focus on a different point in the chain of data collection, analysis, and targeting. Where a group cannot be given control over its data (because there is no structured group with capacity to exercise that control), the goal should be to protect the group's essential interests – primarily, its safety – at the analysis and targeting stages, by anticipating and regulating the riskiest uses of data. Where there is no legal subject to benefit from a privacy right, one solution may be to simply guard against harmful abuses of available data by other stakeholders.

3.3 Affirming and Protecting Group Privacy

As seen above, the distinction between the individual and the group is not always clear-cut – individual privacy and group privacy, while being distinct notions, overlap and affect one another. As a result, two things should go hand in hand to protect group privacy: upgrading individual privacy, and protecting group privacy as such. This must take place in two legal realms, the domestic and international. It must also be effectuated through public and private channels alike. Protecting group privacy should, of course, leverage traditional channels of legislation and treaty-making, but in order to be effective, privacy cannot only be granted in a top-down fashion by lawmakers and through international conventions. Technological solutions must

[48] George Whitecross Paton, *A Textbook of Jurisprudence* 393 (G. W. Paton & David P. Derhamd eds., 4th ed., 1972), cited in *Black's Law Dictionary*, 9th edition, 2009.

also be explored to return a measure of control to data subjects and encourage security, transparency, and accountability. Lastly and crucially, awareness and data literacy must be improved in order for privacy to become more than a relic from the past or a slogan for its advocates, but rather a daily practice of users everywhere.

3.3.1 Through Traditional Levers of Power

Before the spread of the Internet, legal principles and the logistical burdens of the analog world limited the violation of privacy. In recent decades, however, those barriers have been eroded and the application of traditional legal principles in new technological contexts has become uneasy. Whether willingly or accidentally, the tools, resources, and actors that interfere with privacy have multiplied. The imbalance of power between the individual on the one hand and private businesses and governments on the other hand compounds the difficulty of enforcing privacy rights.

But despite assertions of eager private-sector lobbyists or intelligence agency representatives proclaiming the "end of privacy,"[49] national laws and international conventions show there is still universal recognition of the fundamental importance and enduring relevance of privacy; and of the need to safeguard it as a right – even, and especially, in the digital age.

Most democratic countries around the world have privacy frameworks. Some of them are now adjusting those to meet the demands of a globalized and digital world. Most notably, the European Union is currently reforming its regulation and introducing a new, unified E.U.-law; the GDPR pushes for strict data protection compliance for everyone processing data of E.U. persons. Furthermore, the U.S. government initiated an ambitious multi-stakeholder process to develop a Consumer Privacy Bill of Rights; however, for a variety of reasons, progress on this has stalled for now.[50]

Such reforms are important steps towards acknowledging the importance of privacy protection – however, they remain problematic in the sense that they focus more on the collection and transfer of data rather than the type of analysis carried out through that data. In a Big Data world, there are clear limits to the efficiency of notice, choice or consent as tools of data protection.[51] For example, this would be especially clear in the context of discrimination based on personal data: indeed,

---

[49] See Sprenger, Polly. "Sun on Privacy : 'Get Over It.'" Wired, January 26, 1999. http://archive.wired.com/politics/law/news/1999/01/17538; Preston, Alex. "The Death of Privacy." The Guardian, August 3, 2014. http://www.theguardian.com/world/2014/aug/03/internet-death-privacy-google-facebook-alex-preston

[50] Singer, Natasha, 2016. "Why A Push For Online Privacy Is Bogged Down In Washington". New York Times. http://www.nytimes.com/2016/02/29/technology/obamas-effort-on-consumer-privacy-falls-short-critics-say.html?_r=0.

[51] *See* Omer Tene and Jules Polonetsky, Big Data for All: Privacy and User Control in the Age of Analytics, 11 Nw. J. Tech. & Intell. Prop. 239 (2013). http://scholarlycommons.law.northwestern.edu/njtip/vol11/iss5/1, pp. 260–263. *See also* Ohm, "Broken Promises of Privacy: Responding to the Surprising Failure of Anonymization", UCLE Law Review, Vol 57, p. 1763, 2010.

discrimination does not happen in the moment a group member gives his or her consent or when the data is collected, but at a later, undefined point in time, when it is used for targeting.[52]

Although Graham Greenleaf's work and others have noted a fast global expansion in the adoption of privacy laws, in many developing countries, privacy legislations still do not exist, or are very weak, particularly in their enforcement.[53] As a result, advocacy for privacy protection will be essential, both to allow the strengthening of individual privacy in the domestic legislation of all countries and to raise awareness of, and push states to address, the problem of group-related privacy violations.

There has already been some movement, on the international legal plane, on the issue of individual privacy, which advocates for group privacy could build upon. In reaction to the revelation of global surveillance practices, and following a report by former U.N. Commissioner for Human Rights Navi Pillay,[54] the U.N. General Assembly adopted resolution 68/167 on the right to privacy in the digital age in December 2013. In early 2015, the U.N.'s Human Rights Council appointed a Special Rapporteur on the right to privacy.

Group privacy is not explicitly mentioned in any of these documents. However, the new Special Rapporteur has been asked to integrate a gender perspective throughout the work of the mandate. By addressing this issue, the Rapporteur would acknowledge that violations of individual privacy can have a disparate impact on members of certain groups; and that Big Data brings to light a separate interest of group privacy that must be addressed. The Rapporteur's new mandate could thus help build momentum to shed light upon the unique risks posed to minorities in the realm of big data.

With growing awareness of the weak privacy protections in many developing countries, the onus is also on private companies to take a measure of social responsibility and control of their processes. This can be achieved through specific regula-

---

[52] On the predictive nature of Big Data analysis, *see* Kate Crawford and Jason Schultz, Big Data and Due Process: Toward a Framework to Redress Predictive Privacy Harms, 55 B.C.L. Rev. 93 (2014), http://lawdigitalcommons.bc.edu/bclr/vol55/iss1/4 and Ian Kerr and Jessica Earle, "Prediction, Preemption, Presumption: How Big Data Threatens Big Picture Privacy", 66 Stan. L. Rev. Online 65, 2013, http://www.stanfordlawreview.org/online/privacy-and-big-data/prediction-preemption-presumption.

[53] While privacy and data protection laws are generally strong in developed countries, the United Nations Conference on Trade and Development underlines that it remains "inadequate" in other parts of the world. UNCTAD, *Information Economy Report*, 24 March 2015, http://unctad.org/en/PublicationsLibrary/ier2015_en.pdf, pp. 64–65. *E.g.*, at the moment only 14 African 14 countries have or are planning to enact privacy regulations. However, the African Union recently developed a convention on cyber security and personal data protection that would commit member states to establish legal frameworks for e-transactions, protection of data, and punishment of violations. http://www.internetsociety.org/sites/default/files/Internet%20development%20and%20Internet%20governance%20in%20Africa.pdf

[54] *The right to privacy in the digital age*, 30 June 2014, Report of the Office of the United Nations High Commissioner for Human Rights, http://www.ohchr.org/EN/HRBodies/HRC/RegularSessions/Session27/Documents/A.HRC.27.37_en.pdf

tory processes. For example, organizations operating in Europe will soon be compelled to provide reports on non-financial risks and the environmental, social and human rights impacts of their businesses.[55] This is relevant to group protection; and in fact this new regulation concerns public-interest entities, such as banks and insurance companies, that hold a great deal of personal data and whose business practices could therefore interfere with group privacy. Unpacking the mechanisms that affect the privacy and protection of groups would be a first step towards addressing the issue.

Such regulations could become a more generalized practice. Requirements like this could be extended to more industries where group privacy could ostensibly be at stake (for example, data-mining companies), and adopted by a greater number of countries. They could function as an obligation of transparency and accountability, compelling companies to examine their own data processing practices and outline their potential human impact. Going a step further than the requirement to have a publicly available privacy policy, one could imagine a requirement for companies to provide national regulators and private users/clients with an assessment of the potential biases resulting from the company's data processing methods and the ensuing risks for particular groups. Where these risks are too high, or touch upon particularly sensitive issues, national regulators could be given the authority to require modification of the data processing methods so as to minimize their negative effects on vulnerable groups.

Inspiration may also be drawn from existing provisions that already help to protect groups. The current E.U. Data Protection Directive imposes stricter restrictions on the processing of certain categories of sensitive data (health-related, religious, sexual, racial, ethnic, political, etc.), as will its successor, the GDPR. While these restrictions form part of a mechanism to protect individual privacy, they are based on the risk of discrimination and therefore naturally also protect certain vulnerable groups. In order to better address group privacy risks, other legislative texts could emulate and build upon these dispositions. Going one step further they could make explicit their aim of protecting group privacy as well as individual privacy. The incorporation of group privacy concerns into a growing number of laws and regulations around the world would help raise awareness within the companies wishing to do business in the countries concerned and could, in turn, help raise the bar across the world for corporate practices on the protection of vulnerable groups.

Here too, it may appear that group privacy is more effectively protected by regulating data processing and the use of algorithms than by giving people more control over their data.[56] International agreements and "soft law" directly addressing the responsibility of companies in building algorithms and processing data in ethical manners – such as the U.N. Guiding Principles on Business and Human Rights[57] –

[55] http://ec.europa.eu/finance/accounting/non-financial_reporting/index_en.htm
http://ec.europa.eu/finance/accounting/non-financial_reporting/index_en.htm

[56] The regulation of algorithms has already been applied successfully in other areas, such as in the gambling industry.

[57] http://www.ohchr.org/Documents/Publications/GuidingPrinciplesBusinessHR_EN.pdf

would be an additional step towards privacy protection. The Internet Jurisdiction Project[58] is another example of a self-regulatory mechanism for companies operating in a legislative void. In this voluntary alliance, private companies seek consultation from international experts on dilemmas they face regarding freedom of expression. Similar formats could be applied to issues of group privacy protection.

While treaties and legislation are crucial "top-down" instruments to protect privacy in the long run, they also offer limited opportunity to react and adapt at the pace of our fast-changing digital environment. In addition, in the context of Big Data, the discrimination based on algorithms usually does not happen when the data is collected but at a later point in the processing stage. International "soft law" that pushes private actors towards increased transparency about their classification mechanisms and grouping algorithms will therefore be just as important in addressing users' and advocates' concerns.

Companies themselves could participate in developing these instruments, supplementing legislative and regulatory dispositions with their own best practices and industry-wide standards tailored to their specific needs. Privacy laws and regulations must be formulated in broad terms to avoid built-in obsolescence. But corporations' and associations' internal processes and policies are more adaptable to the specificities of their field, and more flexible in the face of fast technological advances. Private actors should therefore participate in the development of a regulatory framework, for example by developing internal corporate rules that specifically outline and limit the use of sensitive information pertaining to groups, at least where they can foresee that the information they collect or the way in which it is processed could be sensitive. They can do so by voluntarily moving towards greater transparency; by adopting group privacy policies that make clear to the public when group-related information is used and how; and by developing compliance processes that allow to regularly control for and correct any violations of that policy. While a changing legislative background would certainly provide part of the incentive for such an evolution, so should the growing public concern for privacy and the notion that privacy is not only good policy, but also sound business.

3.3.2 Through a More Harmonized International Regime for Data Management by Users

As privacy becomes more strongly affirmed in domestic and international law, the avenues for individuals and groups to exercise their privacy rights must become more efficient. Currently, Internet and mobile service users are at the mercy of ill-conceived privacy policies, service providers' compliance with existing laws, and local data protection authorities' strength in enforcing the law. For privacy rights to

[58] http://www.internetjurisdiction.net/

become more meaningful, they must also be exercised more meaningfully by the data subjects themselves.

It is in this view that Greenwood, Pentland, et al. have fleshed out a "New Deal on Data" that includes giving individual citizens "key rights over data that are about them."[59] Their suggestion draws inspiration from the E.U.'s Data Protection Directive, which has, since 1995, successfully altered the practices of major service providers.[60] Going beyond existing laws, they suggest a system of fine-grained individual control over each piece of personal data that would go a long way towards realizing the "informational self-determination" evoked above. In order to achieve this, the authors envision a "trust network" enabled by the alliance of law and technology: on the technological side, all items of data can have "attached labels specifying where the data came from and what they can and cannot be used for."[61] The terms on the labels could, in turn, be matched by the terms of art used in the legal system (in contracts, regulations, etc.) An efficient network would require international harmonization in order for the various labels to be compatible amongst each other and for legal terms to be translated without loss of meaning. While this might seem complex, Greenwood et al. underline that it is akin to Visa Operating Rules and, more generally, the way the credit card network operates.

Their proposal flows naturally from existing systems of privacy protection through user consent, and from the desire to make such consent more fine-grained, more informed, and more genuinely free.[62] Just as it applies to individuals, it could be applied to active and structured groups who benefit from legal personality. One can imagine giving such a group's decision-makers or representatives control over group-specific data: for example, information on the group's inner workings and rules, its culture, and its strategies and plans for the future. However, it cannot protect information that the group is unaware of and which might be extracted from the analysis of members' individual data. For the latter, group protection might once again take a different approach, and focus on restricting analysis of sensitive data categories that are most likely to be used to target or oppress vulnerable groups.

Lastly, while Greenwood et al's approach presents a pragmatic, practical way to hand over data control to users, it comes with new risks of its own. The authors view personal data as a "new asset class," and, in order to encourage positive uses of this data, suggest both "viewing data as money" and creating incentives to share it. While this solves the problem of having massive caches of information siloed within private companies, cordoned off from many potentially beneficial uses, it also risks

---

[59] Daniel Greenwood, Arkadiusz Stopczynski, Brian Sweatt, Thomas Hardjono, and Alex Pentland, "Institutional Controls : the New Deal on Data"

[60] See Google and the right to be forgotten: Silver, Joe. 2014. "Google Must Erase "Inadequate" Links, Court Says". Ars Technica. Accessed April 1 2016. http://arstechnica.com/tech-policy/2014/05/google-must-erase-inadequate-links-court-says/.

[61] Daniel Greenwood, Arkadiusz Stopczynski, Brian Sweatt, Thomas Hardjono, and Alex Pentland, "Institutional Controls : the New Deal on Data"

[62] On the difficulties of "privacy self-management" in the current situation, *see* Daniel J. Solove, "Privacy Self-Management and the Consent Dilemma", 126 Harvard Law Review 1880 (2013), available at http://papers.ssrn.com/sol3/papers.cfm?abstract_id=2171018.

increasing inequalities. If services or financial incentives are provided in exchange for personal data, it is easy to imagine that the most vulnerable elements in society – whether individuals or groups – will be more eager to part with their information, thereby making privacy a privilege. This can be fought, at least in part, by improving data literacy and privacy awareness, as discussed in subsection 3.3.4.

3.3.3 By Improving Data Security and Accountability for Breaches

As law and technology evolve to ensure greater user control of individual and group data, technology can also afford us stricter control of the downstream uses made of such data. This is perhaps the most promising avenue for protecting group privacy in the case of passive groups or even in the case where data analysis can extract information that the group would not know how to protect. Accountability should not be left to the whim of governments' and businesses' goodwill alone; alongside stronger laws and regulations, technology itself can help protect both individual and group privacy.

Greenwood et al. thus suggest integrating "computational law technology" into personal data systems in order to automatically verify the terms of use agreed upon for each item of data and the compliance of parties to these terms. Similarly, de Montjoye et al. propose a system dubbed openPDS, "an open-source Personal Data Store enabling the user to collect, store, and give access to their data while protecting their privacy."[63] The system ensures that most processing of a user's personal data does not take place on a third-party server, but rather locally within the user's own personal data store, a secured digital space under his control. This allows service providers and applications to "see" the user's data when needed, without its being handed over to their control. PDSs can also engage in "group computation" to answer specific questions about groups of people, without service providers and other actors gaining access to or ownership of the data of the people or group thus concerned.

Technology can also serve to enhance accountability. For example, data analysis giant Palantir's software, used by U.S. government agencies and increasingly also private financial institutions, purportedly features "baked in" privacy protective capabilities. Its Gotham platform tracks any use of the data, with each action being attributed, and stored[64]; its intelligence software also features "immutable audit functions."[65] However, while this "audit-trail technology" is reportedly built into the private-sector versions of Palantir software, its use is not mandated, and the com-

[63] Yves-Alexandre de Montjoye, Samuel S. Wang, Alex Pentland, "On the Trusted Use of Large-Scale Personal Data", 2012.

[64] Palantir Gotham Overview, https://www.palantir.com/palantir-gotham/.

[65] Palantir Intelligence, https://www.palantir.com/solutions/intelligence/.

pany itself has admitted that it cannot control how its clients use its products.[66] In addition, while this technology allows to track unauthorized use or tampering with the data, it does not incorporate additional privacy protections extraneous to the company's internal rules. It simply creates an indelible "trail" that facilitates audits. Palantir is precisely the kind of company that has drawn fierce criticism from civil rights advocates for enabling mass surveillance.[67] Yet its use of "civil liberties engineers" and its inclusion of audit features in its software can still point us towards a solution, and one that is particularly relevant to the protection of group privacy. Where individual controls are insufficient, they can be supplemented by audit technologies that help maintain a clear chain of responsibility in case of abuse. Here too, of course, technology, law, and policy must go hand in hand: technology can improve accountability only if the will or obligation exists to carry out regular audits and to sanction violations.

3.3.4 Through Improving Awareness and Data Literacy

Currently, Big Data appears to most citizens to be too complex, too blurry a notion, and too technical to be readily grasped. To ensure that the yet-to-be-defined path of Big Data leads to societal welfare and prosperity, people must be empowered to engage in a much-needed debate about what kind of data-driven world we want, ensuring that they have the capacities to act as mature citizens and shapers of a digital world.

First, education is crucial for the obvious reason that only people who know how something works are able to shape it. In the age of Big Data, the wider public needs to better understand the basics of digital technology, data science and algorithms. Not every schoolchild should become a data scientist, but basic technical skills such as the essentials of programming could enable us to better understand our new digital environment; and hence to help open the "black box" and demystify technology.

This also means discussing the societal and ethical implications of Big Data at an early stage, including its potential downsides, and explaining how this can impact a person's life and the lives of others in the analog world – for example, by making clear the possibility of algorithm-based discrimination on traits that would have been invisible in the analog world (such as religion or sexual orientation).

Secondly, this means investing in university training to point young people towards new careers. We are just at the beginning of the Big Data era, and alongside the increasing demand for data scientists, we can assume that in the future more experts in data law, data ethics, privacy and digital rights will be needed.

---

[66] Quentin Hardy, "Unlocking Secrets, If Not Its Own Values", *The New York Times*, May 31, 2014, http://www.nytimes.com/2014/06/01/business/unlocking-secrets-if-not-its-own-value.html?

[67] http://www.bloomberg.com/bw/magazine/palantir-the-vanguard-of-cyberterror-security-11222011.html

Thirdly, programmers, data experts and the like are exerting increasing influence on our daily lives as digital technologies and the accumulation of huge quantities of data affect us all. Many of them might not even be aware of how their work might harm others, such as minorities or other members of other particular groups. The responsibility of each individual should be discussed when producing algorithms and other digital tools. Ethical training and guidelines – as we have seen take root in other industries – are urgently needed for these experts.

Fourth and lastly, civil society organizations must be involved in the discussion. This includes raising awareness among digital rights advocates on the potential negative implications for certain groups, and capacity-building for NGOs that advocate for groups and minorities in the analog world. More campaigning needs to be done to expose not only the potential risks of Big Data for individuals but also, and more importantly, the impact these potential risks may have on groups.

3.4 Conclusion

Big Data has blurred the boundaries between individual and group data. Through the sheer number and richness of databases and the increasing sophistication of algorithms, the "breadcrumbs" left behind by each one of us have not only multiplied to a degree that calls our individual privacy into question; they have also created new risks for groups, the members of which can be targeted and discriminated against unbeknownst to themselves, or even unbeknownst to data analysts. This prompts us to enrich our understanding of privacy. Where individual privacy might once have sufficed to rein in state and corporate surveillance practices as well as the neighbors' curiosity, and sufficed to give individuals a measure of control over their reputations and security, today it can leave groups vulnerable to discrimination and targeting and, what's more, leave them unaware of that risk. The concept of group privacy attempts to supplement individual privacy by addressing this blindspot.

Group privacy is not, however, without complications of its own. Indeed, creating a simple, one-dimensional group privacy right is no silver bullet: such a right can only provide effective protection where there is a group possessed of legal personality able to enforce it before a (domestic or international) court or tribunal. Yet Big Data's particularity lies precisely in its ability to extract valuable information about passive groups with no such self-awareness or capacity. Thus, on the one hand, a group privacy right can help active, structured groups assert their informational self-determination and protect their own interests. On the other hand, it must be supplemented by additional protections that recognize and address the privacy interest of passive groups extracted at the data analysis stage.

This points us towards a multi-pronged approach to strengthen the protection of privacy. Traditional avenues, including conventions on the international plane and legislation in the domestic legal sphere, are indispensable to reaffirm the importance of privacy and further public debate about its application to groups. These should not focus only on setting the conditions for lawful data collection, but also on limiting and sanctioning the risky downstream potential uses of such data.

The introduction of harmonized regulation on data sharing could also afford users a greater measure of control over their own data and increase transparency surrounding the ways in which our myriad "breadcrumbs" of information are used. At the same time, the private sector must be harnessed – both to help develop technology that ensures greater accountability for privacy breaches, and to encourage the social responsibility of businesses where local privacy laws are weak.

Lastly, none of these changes can have a meaningful impact without increased data literacy across the board, so that individuals become more aware of the impact of their actions not only on their own safety, but also on that of others. Improving privacy protections is not an impediment to the myriad potentialities of Big Data – but rather the condition for this potential to be unleashed in a responsible and socially beneficial way.

Bibliography

"Guiding Principles on Business and Human Rights." *United Nations Human Rights Office of the High Commissioner*. HR/PUB/11/04 (2011). http://www.ohchr.org/Documents/Publications/GuidingPrinciplesBusinessHR_EN.pdf.

"Intelligence." *Palantir Technologies*. 2016. https://www.palantir.com/solutions/intelligence/.

"Internet & Jurisdiction Project—A Global Muti-Stakeholder Dialogue Process." *Internetjurisdiction.net*http://www.internetjurisdiction.net/.

"Non-Financial Reporting- European Commission." *Ec.Europa.Eu* January 15, 2016. http://ec.europa.eu/finance/accounting/non-financial_reporting/index_en.htm.

"Palantir Gotham." *Palantir Technologies* 2016. https://www.palantir.com/palantir-gotham/.

"Riding With the Stars: Passenger Privacy in the NYC Taxicab Dataset." *Neustar Research*, September 15, 2014. http://research.neustar.biz/2014/09/15/riding-with-the-stars-passenger-privacy-in-the-nyc-taxicab-dataset/.

"Using Cell Phone Data to Curb the Spread of Malaria." *Harvard T.H. Chan School of Public Health News*, October 11, 2012, http://www.hsph.harvard.edu/news/press-releases/cell-phone-data-malaria/.

Akinsuyi, F. Franklin. 2015. "Data protection and privacy laws Nigeria, a trillion dollar opportunity!!" Social Science Research Network. April 24, 2015. http://ssrn.com/abstract=2598603.

Bengtsson, Linus, et al. 2011. Improved response to disasters and outbreaks by tracking population movements with mobile phone network data: A post-earthquake geospatial study in Haiti. *PLoS Med* 8(8).

Berlee, Anna. 2015 "Using NYC taxi data to identify muslim taxi drivers." *The Interdisciplinary Internet Institute.* January 21, 2015. http://www.theiii.org/index.php/997/using-nyc-taxi-data-to-identify-muslim-taxi-drivers/.

Bloustein, Edward J. 1964. Privacy as an aspect of human dignity: An answer to Dean Prosser. *New York University Law Review*, 962–1007.

Blumenstock, Joshua, et al. 2015. Neighborhood and network segregation: Ethnic homophily in a silently separate society. *Proc. NetMob*.

Cassese, Antonio. 1995. *Self-determination of peoples: A legal reappraisal*. Cambridge: Cambridge University Press, p. 11.

Charney, Jonathan I. 2001. Self-determination: Chechnya, Kosovo, and East Timor. *Vanderbilt Journal of Transnational Law* 34: 455.

Crawford, Kate, and Jason Schultz. 2014. Big data and due process: Toward a framework to redress predictive privacy harms. 55 *B.C.L. Rev*. 93, http://lawdigitalcommons.bc.edu/bclr/vol55/iss1/4.

de Corbion, Alexandrine Pirlot. Lebanon: It's time to you're your international position on privacy into action at the national level. *Privacy International*, May 26. 2015, https://www.privacyinternational.org/node/586.

de Montjoye, Yves-Alexandre, Samuel S. Wang and Alex Pentland. 2012. On the trusted use of large-scale personal data. *IEEE Data Engineering Bulletin*, 35–4.

Decuyper, Adeline, et al. 2014 Estimating food consumption and poverty indices with mobile phone data. *arXiv preprint arXiv:1412.2595*.

European Parliament and Council. *Directive on the Protection of Individuals with Regard to the Processing of Personal Data and on the Free Movement of Such Data*. (95/46/EC) October 24, 1995. http://ec.europa.eu/justice/policies/privacy/docs/95-46-ce/dir1995-46_part1_en.pdf.

G.A. Resolution 1803, U.N. GAOR, 17th session, Supp. No. 17, at 15, U.N. Doc A/5217 1962.

Greenwood, Daniel, et al. 2014. The new deal on data: A framework for institutional controls. In *Privacy, big data, and the pubic good*, ed. Julia Lane, Victoria Stodden, Stefan Bender, and Helen Nissenbaum, 192–210. Cambridge: Cambridge University Press.

Hardy, Quentin. 2014. "Unlocking secrets, if not its own values," *The New York Times*, May 31, 2014, http://www.nytimes.com/2014/06/01/business/unlocking-secrets-if-not-its-own-value.html?.

Kerr, Ian, and Jessica Earle. 2013. Prediction, preemption, presumption: How big data threatens big picture privacy. 66 *Stan. L. Rev*. 65, http://www.stanfordlawreview.org/online/privacy-and-big-data/prediction-preemption-presumption.

Livingston, Steven, and Gregor Walter-Drop (eds.). 2013. *Bits and atoms: Information and communication technology in areas of limited statehood*. New York: Oxford University Press.

Mao, Huina, et al. 2013. Mobile communications reveal the regional economy in Côte d'Ivoire. *Proc. of NetMob*.

Mayer-Schönberger, Viktor, and Kenneth Cukier. 2013. *Big data: A revolution that will transform how we live, work, and think*. London: John Murray.

Nissenbaum, H., and F. Brunton. 2015. *Obfuscation: A user's guide for privacy and protest*. Cambridge: MIT Press.

Nissenbaum, Helen. 2010. *Privacy in context: Technology, policy, and the integrity of social life*. Palo Alto, CA: Stanford University Press. Spanish Translation *Privacidad Amenazada: Tecnología, Política y la Integridad de la Vida Social*. Mexico City: Océano, 2011.

Nissenbaum, H., and M. Price (eds.). 2004. *Academy and the internet*. New York: Peter Lang Publishing Company.

Nyirenda-Jere, Towela, and Tesfaye Biru. "Internet Development and Internet Governance in Africa." *Internet Society*. May 22, 2015. http://www.internetsociety.org/sites/default/files/Internet%20development%20and%20Internet%20governance%20in%20Africa.pdf.

Ohm, Paul. 2010. Broken promises of privacy: Responding to the surprising failure of anonymization. *UCLE Law Review* 57: 1763.

Omer, Tene, and Jules Polonetsky.2013. Big data for all: Privacy and user control in the age of analytics. *11 Nw. J. Tech. & Intell. Prop*. 239: 260–263. http://scholarlycommons.law.northwestern.edu/njtip/vol11/iss5/1.

Paton, George Whitecross. 1972. *A textbook of jurisprudence* 393 (G. W. Paton & David P. Derhamd eds., 4th ed., cited in *Black's law dictionary*, 9th edition, 2009.

Posner, Richard A. The economics of privacy. *The American Economic Review*, 71(2): 505–409. http://www.jstor.org/stable/1815754.

Preston, Alex. 2014. "The Death of Privacy." *The Guardian*, August 3, 2014. http://www.theguardian.com/world/2014/aug/03/internet-death-privacy-google-facebook-alex-preston.

Raymond, Nathaniel. "Beyond 'Do no harm' and individual consent: Reckoning with the emerging ethical challenges of civil society's use of data." *(forthcoming)*

Rosenfeld, Friedrich. 2010. "Collective reparation for victims of armed conflict," 92 *International Review of the Red Cross* 731.

Rotenberg, Mark, Julia Horwitz, and Jeremie Scott (eds.). 2015. *Privacy in the modern age: The search for solutions*. New York: The New Press.

Smith, Christopher, Afra Mashhadi, and Licia Capra. 2013. "Ubiquitous sensing for mapping poverty in developing countries." *Paper submitted to the Orange D4D Challenge*.

Singer, Natasha. 2016. Why a push for online privacy is bogged down in Washington. *New York Times*. http://www.nytimes.com/2016/02/29/technology/obamas-effort-on-consumer-privacy--falls-short-critics-say.html?_r=0.

Solove, Daniel J. 2013. Privacy self-management and the consent dilemma. 126 *Harvard Law Review* 1880. http://papers.ssrn.com/sol3/papers.cfm?abstract_id=2171018.

Solove, Daniel J. 2008. "Understanding Privacy." GWU Law School Public Law Research Paper No.420. Harvard University Press. http://papers.ssrn.com/sol3/papers.cfm?abstract_id=1127888.

Solove, Daniel. 2002. Conceptualizing privacy. *90 California Law Review* 1087: 1089.

Sprenger, Polly.1999. Sun on privacy: 'Get over it.' *Wired*, January 26, 1999. http://archive.wired.com/politics/law/news/1999/01/17538.

Tatem, Andrew J., et al. 2014. Integrating rapid risk mapping and mobile phone call record data for strategic malaria elimination planning. *Malaria Journal* 13(1): 1–16.

Theriault, Denis C. 2015. Black lives matter: Oregon justice department searched social media hashtags. *Oregonlive.Com*, November 10, 2015. http://www.oregonlive.com/politics/index.ssf/2015/11/black_lives_matter_oregon_just.html.

U.N. General Assembly, *Convention Relating to the Status of Refugees*. July 28, 1951. http://www.unhcr.org/3b66c2aa10.html.

U.N. General Assembly, *International Covenant on Civil and Political Rights*, G.A. res. 2200A (XXI), December 16, 1966, 21 U.N. GAOR Supp. (No.16) at 52, U.N. Doc. A/6316 (1966), 999 U.N.T.S. 171, *entered into force* March 23, 1976. http://www.ohchr.org/en/professionalinterest/pages/ccpr.aspx.

U.N. Human Rights Council, 28th Session. Resolution. *The Right to Privacy in the Digital Age*. (A/HRC/RES/28/16) April 1, 2015. http://ap.ohchr.org/documents/dpage_e.aspx?si=A/HRC/RES/28/16.

U.N. Human Rights Council. 27th Session. Report of the Office of the United Nations High Commissioner for Human Rights. *The Right to Privacy in the Digital Age*. (A/HRC/27/37). June 30, 2014. http://www.ohchr.org/EN/HRBodies/HRC/RegularSessions/Session27/Documents/A.HRC.27.37_en.pdf.

United Nations Conference on Trade and Development (UNCTAD), *Information Economy Report*, 24 March 2015, http://unctad.org/en/PublicationsLibrary/ier2015_en.pdf, pp. 64–65.*E.g*.

Van den Hoven, Jeroen. 1997. Computer ethics and moral methodology. *Metaphilosophy* 28(3): 234–248.

Vance, Ashlee, and Brad Stone. 2011. "Palantir, the War on Terror's Secret Weapon." *Bloomberg.com*, December 22, 2011. http://www.bloomberg.com/bw/magazine/palantir-the-vanguard-of-cyberterror-security-11222011.html.

Warren, Samuel D., and Louis D. Brandeis. 1890. The right to privacy. *Harvard Law Review*, IV(5).

Westin, Alan F. 1968. Privacy and freedom. *25 Washington and Lee Law Review*. 166. http://scholarlycommons.law.wlu.edu/wlulr/vol25/iss1/20.

Whitman, James Q. 2004. Two western cultures of privacy: Dignity versus liberty. 113 *Yale Law Journal* 1153.

Whong, Chris "Foiling NYC'S Taxi Trip Data" *Chriswhong.Com*. 2016. http://chriswhong.com/open-data/foil_nyc_taxi/.

Williams, Apryl, and Doris Domoszlai. "#Blacktwitter: A Networked Cultural Identity | Harmony Institute". *Harmony-Institute.Org*, August 6, 2016. http://harmony-institute.org/latest/2013/08/06/blacktwitter-a-networked-cultural-identity/.

Chapter 4
Beyond "Do No Harm" and Individual Consent: Reckoning with the Emerging Ethical Challenges of Civil Society's Use of Data

Nathaniel A. Raymond

Abstract The digital revolution is transforming how governments, the private sector, and civil society view the possibilities and perils inherent in the use of new Information Communication Technologies (ICTs). For humanitarian, human rights, and development actors, well founded anxieties are arising about the uncharted and poorly defined ethical implications of these increasingly commonplace tools and tactics. Unethical and potentially illegal "disaster experimentation" will continue to occur as long as the current gap in ethical doctrine for the use of these technologies persists. This chapter explores two critical ethical "blindspots" related to the current use of ICTs by civil society actors – the increasing critical importance of demographically identifiable information and the deployment of remote data collection strategies when individual informed consent is not possible.

Keywords Demographically identifiable information • Humanitarian innovation • Human rights • Data ethics • Conflict • Satellite data • Accountability • Civil society organisations

4.1 Introduction: New Technologies and New Ethical "Blindspots"

The digital revolution is transforming how governments, the private sector, and civil society view the possibilities and perils inherent in the use of new Information Communication Technologies (ICTs).[1] For humanitarian, human rights, and development actors, the initial excitement which accompanied the emergence of crowd

[1] For a definition of ICTs, please see: https://www.cs.cmu.edu/~rtongia/ICT4SD_Ch_2DOUBLEHYPHENICT.pdf

N.A. Raymond (✉)
Signal Program on Human Security and Technology, Harvard University, Cambridge, MA 02138, USA
e-mail: nathaniel.raymond@gmail.com

L. Taylor et al. (eds.), *Group Privacy*, Philosophical Studies Series 126,
DOI 10.1007/978-3-319-46608-8_4

mapping platforms such as Ushahidi and the increased availability of satellite imagery has now begun to give way to well founded anxieties about the uncharted and poorly defined ethical implications of these increasingly commonplace tools and tactics.

One needs look no further than Sean Martin McDonald's landmark 2016 paper, *Ebola: A Big Data Disaster*, to see that these concerns are increasingly being born out by recent use cases of ICTs in the context of complex disasters. McDonald makes the case that the use of ICTs to capture Call Detail Records (CDRs) during the 2014–2015 West African Ebola Pandemic violated local and international legal standards; infringed on the individual and group privacy protections of civilian populations; and employed these tools towards achieving a still largely ill-defined technical and operational goal.

It is unfortunately highly likely that the type of clearly unethical and potentially illegal "disaster experimentation" that McDonald documents by civil society actors in the context of the ebola crisis will continue to occur as long as the current gap in ethical doctrine for the use of these technologies persists. In the absence of any substantive outside legal or regulatory enforcement either domestically or internationally for civil society's use of ICTs and the data they produce, the development of ethical norms that may encourage self-regulation by civil society groups becomes essential.

This chapter aims to identify, define, and explore two critical ethical "blindspots" related to the current use of ICTs by civil society actors – the increasing critical importance of demographically identifiable information and the deployment of remote data collection strategies when individual informed consent is not possible. This chapter contends that these two largely unaddressed blindspots, in particular, are preventing civil society from effectively responding to the new ethical challenges unique to the emerging use of ICTs.

4.2 The Ethical Doctrine Gap in Civil Society's Current Use of ICTs and Data

The rapid and ongoing adoption and adaption of information communication technologies (ICT) within the past decade by humanitarian and human rights actors for the purposes of capturing and analyzing multiple forms of digital data is a significant turning point in the history of civil society. While many types of organizations may comprise what can be defined as "civil society", this chapter focuses primarily on non-governmental organizations (NGOs) that are engaged in the provision of humanitarian assistance, evidence collection for humanitarian advocacy and accountability purposes, and community-based organizations engaged in development and peace building activities.

ICTs now provide these non-governmental organizations sense making, outreach, and situational awareness capacities that, when they have been available to others in the past, were primarily the domain of private sector and governmental actors, particularly militaries and intelligence services. My colleagues Brittany Card, Ziad al Achkar, and I identify in a 2015 article for the European Interagency Security Forum three common uses cases of ICTs specific to the humanitarian sector.

These use cases provide clear examples of how many civil society groups, not only humanitarian organizations, are applying these increasingly commonplace tools for a diverse range of constantly evolving purposes:

- Remotely collecting and analysing social media, geospatial data and other sources of data;
- Communicating information in order to improve situational awareness and dispel rumours; and
- Connecting affected populations to response activities.

While the potential benefits of these applications of ICTs and the data derived from them for civil society groups may appear obvious, the unique and emerging ethical challenges that these technologies and their applications may create and/or magnify are significantly less clear. Brittany Card and I, in our white paper *Applying Humanitarian Principles to Current Uses of Information Communication Technologies: Gaps in Doctrine, Challenges to Practice*, conclude that there is a general lack of "minimum standards" for the provision and use of ICTs in humanitarian action.

4.3 The Emergence of Ad Hoc Codes of Conduct

NGOs, in particular voluntary technical organizations (VTOs), have nonetheless begun to generate and adopt ICT application-specific ethics codes as they seek to face these challenges despite clear consensus about how to address these glaring gaps in humanitarian ethical doctrine. While a far cry from minimum standards, these initial attempts at providing organizations ethical guidance currently represent the state of the art.

These ethical regimes, often in the form of individual VTO "codes of conduct", are emerging simultaneous to concerns being raised in the literature about the potential implications these new, little understood ethical dynamics inherent in civil society applications of ICTs may have for vulnerable populations. My colleagues Caitlin Howarth, Jonathan Hutson, and I write in *Crisis Mapping Needs an Ethical Compass*, that civil society groups employing ICTs:

> ...often risk inadvertently creating new perils for those whom they strive to help. As such, some of the pressing questions facing the field of crisis mapping that have yet to be answered in a generalizable way include: What information should be shown publicly, and when and how should it be shown? When should it not be shown? Do crisis mappers sometimes unintentionally provide bad actors with very useful intelligence? Are at-risk populations endangered by sharing information with crisis mapping initiatives and/or social media – even when this is done remotely and with the use of encryption? What happens to vulnerable civilians if crisis mapping data is wrong? What happens to them if the data is right? What

responsibility does the crisis mapping community have to report and share mistakes transparently? If crisis mappers are the first to spot an emerging threat, then what is the most ethical and effective way to alert people on the ground who may be in imminent danger? How can sensitive data be kept more secure from hackers? When is the level of risk to vulnerable populations – or to the crisis mapper – too high to engage in crisis mapping? Who is ultimately accountable for measuring, evaluating and mitigating these risks?

Even without an emerging consensus to many of these critical questions, a nascent ethical regime animated by apparently shared, cross-cutting concerns can seen to be emerging primarily through these ad hoc codes of conduct being promulgated by VTOs and some larger NGOs. In most cases, these VTO-driven codes of conduct are strongly influenced by two particular ethical antecedents:

1. The *primum non nocere* ("above all – do no harm") concept found in medical and social science research ethics; and
2. The standards embodied by the Red Cross/NGO Code of Conduct.[2]

While other sources of tradition may influence them to varying degrees, a general literature review suggests that these two antecedents appear to be the most influential. The intent of these codes of conduct, and the sources of ethical tradition upon which they are largely based, are both consistent with the past practices of other, related fields of data collection and analysis. They also incorporate, in many cases verbatim, the core values that define humanitarian action, such as "humanity", "impartiality", "dignity", and "neutrality", which these applications of ICTs often aspire to uphold and advance.

Additionally, these ethical regimes often, understandably, reference or draw upon the primary operational guidance available to date for conducting this work, the International Committee of the Red Cross' *Professional Standards for Protection Work Carried Out by Humanitarian and Human Rights Actors in Armed Conflict and Other Situations of Violence* (hereafter, "Professional Standards for Protection Work").[3] However, these ethical regimes are fundamentally insufficient for addressing the unique ethical challenges of current civil society applications of ICTs and the data derived from them.

These "Do No Harm" and humanitarian principle-based regimes, this chapter argues, are severely insufficient to meeting the ethical challenges of the networked age by themselves. These concepts alone are inadequate to the present historical moment because they neither countenance nor address what this chapter identifies as the two primary requirements for any comprehensive ethical regime that seeks to guide civil society's burgeoning application of ICTs. These requirements are being able to define and address the heretofore undefined concepts of "Demographically Identifiable Data", or DII, and what this chapter refers to as the "Consent Paradox".

This chapter will seek to define both the concepts of DII and the Consent Paradox, exploring their current and potential implications to vulnerable populations from

[2] Code of Conduct. IFRC. Retrieved from http://www.ifrc.org/en/publications-and-reports/code-of-conduct/

[3] http://www.ifrc.org/en/who-we-are/vision-and-mission/the-seven-fundamental-principles/

civil society’s applications of ICTs. Additionally, this chapter will also seek to understand the roadblocks these specific challenges present to creating comprehensive, coherent, and realistic ethical regimes for this sector’s use of ICT-derived data, as well as potential approaches for helping address them.

4.4 Defining What Constitutes “Beneficence” and “Non-maleficence” in Current Practice

The ethical state of the art, so to speak, of civil society’s emerging applications of ICTs and the data derived from them must first be more fully understood before the concepts of DII and the Consent Paradox, as well as their potential implications, can be appropriately and fully defined. At present, two main objectives appear to broadly underpin the emerging ethical norms guiding ICT application by civil society.

First, there is the clear aim of avoiding the unintentional infliction of harm upon the populations which those applying ICT and their data seek to serve through the use of these technologies and the resulting data they may provide. Secondly, these groups aspire to apply ICTs and their data in a way that upholds the defining values of the humanitarian and human rights fields – of which many ICT data users self-identify as members. Chief and most cited amongst these values is the principle of “humanity” (or some derivation of it), which is defined by the International Federation of Red Cross and Red Crescent Societies, in part, as the protection “of life and health” and the ensuring “of respect for the human being”.

In the vocabulary of medical ethics, from which some of these ethical concepts originate, these two goals express what is referred to as the ethical principles of “non-maleficence” and “beneficence”, respectively. The University of California San Francisco (UCSF) Medical School defines non-maleficence as follows:

> Non-maleficence means to “do no harm.” Physicians must refrain from providing ineffective treatments or acting with malice toward patients…The pertinent ethical issue is whether the benefits outweigh the burdens.

Beneficence, which can often be confused with non-maleficence, is defined by UCSF Medical School below:

> Beneficence is action that is done for the benefit of others. Beneficent actions can be taken to help prevent or remove harms or to simply improve the situation of others.

Exploring current civil society conceptions of non-maleficence, as opposed to the comparatively more easily applied concept of beneficence, is essential for assessing whether the current ethical state of the art in civil society’s use of data is sufficient to address the field-specific challenges it faces. In the context of this chapter, definitions of engaging in the ethic of non-maleficence is limited to the following: The ability to ethically weigh the balance of consequences related to any potential intervention and the ability to ensure the professional competencies necessary to engage in beneficence as defined by current practice in the field.

As this chapter will show, civil society groups, based on the definition provided above, are at present largely incapable of engaging in non-maleficence. This is the case because the primary dangers that their emerging ethical regimes attempt to address are increasingly anachronistic in the face of the threats inherent in current applications of ICTs.

4.5 Current Definitions of Non-maleficence: Protecting "PII" and Obtaining Individual Informed Consent

What the field perceives the current ethical challenges it faces to be that could result in maleficence, or the doing of harm, must be first identified to begin to assess whether its definitions of non-maleficence are sufficient. Two clear and interrelated mechanisms by which harm could be potentially inflicted to the populations that civil society actors applying ICTs and their data appear to routinely be identified as major issues driving the creation of ethical regimes for this sector.

One, is the disclosure of "personally identifiable information", or PII, collected by data-driven interventions. A definition of PII provided by the United States Department of Labor (n.d.) is helpful to understanding the myriad of potential data that can constitute PII:

> Any representation of information that permits the identity of an individual to whom the information applies to be reasonably inferred by either direct or indirect means. Further, PII is defined as information: (i) that directly identifies an individual (e.g., name, address, social security number or other identifying number or code, telephone number, email address, etc.) or (ii) by which an agency intends to identify specific individuals in conjunction with other data elements, i.e., indirect identification. (These data elements may include a combination of gender, race, birth date, geographic indicator, and other descriptors). Additionally, information permitting the physical or online contacting of a specific individual is the same as personally identifiable information. This information can be maintained in either paper, electronic or other media.

The threats that the inadvertent and/or unauthorized leak, intercept, or theft of PII may pose in the use of data in humanitarian and human rights contexts are multitudinous, complex, and are, unfortunately, increasingly well documented. Chamales and Baker's *Securing Crisis Maps in Conflict Zones* (2011) presents several examples, including recent incidents in Sudan and Pakistan, where PII and DII disclosure by civil society data deployments put civilians and practitioners at potential risk of harm.

Some of these threats that may result from PII collected by a data collection deployment being compromised can include the exposure of specific individuals and/or populations to societal stigma; targeting of individuals and groups for violent reprisal; subjecting undocumented populations to law enforcement and/or deportation proceedings; and increased risk of exploitation by various actors.

The second mechanism that civil society has generally acknowledged as a reason for creating ethical regimes governing data deployments involving vulnerable populations is the issue of the collection of PII without the informed consent of those providing it. This issue is intricately and intimately linked to the threat of improper disclosure of PII discussed above.

While there is no accepted "humanitarian" or "human rights" field-specific definition informed consent, per say, the following three part definition of the concept from the US Department of Health and Human Services (n.d.) pertaining to human subjects research generally defines the idea:

1. disclosing to potential research subjects information needed to make an informed decision;
2. facilitating the understanding of what has been disclosed; and
3. promoting the voluntariness of the decision about whether or not to participate in the research.

In both governmental and non-governmental definitions, the ability for an individual whose PII is being collected to consent to participating in its collection, and to do so with an informed understanding of how it will be used, is consistently connected to the often interchangeable values of upholding their "humanity" and "dignity". An example of this concept of informed consent in practice by a civil society actor engaged in data collection from vulnerable populations is the data policy of Oxfam Great Britain (2015), who state:

> Participants have the right to be fully informed in order to make a decision about their participation in any data activity...Oxfam and its agents will gain informed and voluntary consent before obtaining any information from participants. Data will only be used for the purpose it was collected for.

4.6 Single Stream Ethics Are Insufficient in a Multi-stream World

The field's understandable aspirational focus on the protection of PII and the receipt of informed consent for data collection and use, as evidenced above, is admirable and well intentioned. However, what this chapter contends is that the attention paid to these two conjoined issues as critical aspects of achieving the ethic of non-maleficence fails to address two critical operational realities:

1. The evolving nature of digital data precludes fully eliminating actionable PII from data streams, and in fact, can make even de-identified data actionable on a large scale and in unanticipated ways; and
2. When, why, and how data is increasingly collected by civil society ICT deployments increasingly precludes the ability to obtain individual informed consent.

As evidence of the inability to fully de-identify PII in digital media, de Montjoye, Kendall, and Kerry state in *Enabling Humanitarian Use of Mobile Phone Data (2014)*, that:

> ...elimination of specific identifiers is not enough to prevent re-identification. The anonymity of such datasets has been compromised before and research shows that, in mobile phone datasets, knowing as few as four data points—approximate places and times where an individual was when they made a call or send a text—is enough to re-identify 95 % of people in a given dataset. In general, there will be very few people who are in the same

> place at the same time on four different occasions, which creates a unique "signature" for the individual making it easy to isolate them as unique in the dataset. The same research also used unicity to shows that simply anonymized mobile phone datasets provide little anonymity even when coarsened or noised.

The ability to reconstruct individual and group identities from de-identified mobile data sets raises the issue of the threat from both these and other sources of data to vulnerable populations not simply being individual in nature but demographic as well. As Chamales and Baker (2011) note, information that could be accessed by potential perpetrators of abuses based on data harvested by civil society data deployments can also include what this chapter describes as "demographically identifiable information" (DII), rather than simply the initial individually identifiable PII data alone. They state:

> Hostile organizations such as oppressive governments do not necessarily need a reason to target a specific individual or group, however individuals who report on the activities by these organizations can make themselves a target for attack and retribution. In a conflict, those reporters may be citizens communicating over social media, submitting text messages to a crisis mapping platform, or professional journalists. The information related by those reports can also be used to identify vulnerable groups such as refugees, those acting against the hostile powers, or response organizations – as was the case with the Taliban's threat to target foreign aid workers responding to the 2010 floods in Pakistan.

The Red Cross Professional Standards for Protection Work (2013) also recognizes the threat presented by emerging and increasing commonplace digital approaches to information collection, sharing, and aggregation, stating:

> The protection of the sources of information that might decide to use electronic means (blog, SMS, email, tweets, social networks, etc.) to rapidly communicate information to the public, or to third parties, while unaware of the risks of being identified or tracked by the authorities or armed groups who might take actions against them. In some cases, retaliation might affect a whole community.

What is more, this dynamic is complicated by the fact that the diversity of data types and sources being used by ICT deploying organizations are increasingly heterogeneous and changing. Organizations are now increasingly using the combination, or "fusion", of multiple streams together to create humanitarian and human rights data products. The paradigm of individually consented, single source streams of traditional PII data is becoming eclipsed by this integration of multiple data streams together into an aggregated, derived product.

Thus, even if some of data was originally obtained through consent, some initially consented single source streams of data are likely being used to develop cross-corroborated insights that may significantly transcend the initial stated purposes for which one or more stream of data was first collected. This act of fusion invalidates any previous informed consent specific to a single stream's collection if the terms of the consent did not cover its integration with other streams of data.

While humanitarian standards generally call for seeking informed consent as a core component of all aid activities, there exists no accepted structure for tracking informed consent in either single streams of civil society data or the more and more prevalent "multi-stream" approach. It is in these multi-stream approaches, which are increasingly the norm, that DII can both be generated and can become dangerous.

What's more, these "multi-stream" approaches appear to be becoming more prevalent precisely because DII, rather than PII, is more and more often the goal of these deployments. Often these deployments are also being necessitated by the fact that individual and informed consent-based collection of data is impossible in non-permissive environments where these types of demographic insights are most required by responding agencies.

4.7 Defining DII and Understanding Its Potential Operational Implications

Understanding the potential threat of DII and the current gap in available ethical guidance to civil society groups related to it requires attempting to define it. An online literature review found less than a dozen uses of the term "demographically identifiable information", and thus, no accepted definition of the term outside the one provided below by the author appears available.

While there is some mention of "demographic information", it is often presented as a subset of personal identifiable information, such as name, age, ethnicity, etc. This chapter contends that DII requires its own category and science of identifiable data specific to itself.

This absence of a clearly articulated concept of DII is striking given its critical role in now common digital, networked data collection approaches, such as smartphone apps, social media, and any crowd-sourced platform offered by the private sector. The lack of a standard definition of this term is itself evidence of the enormity of the technical and doctrinal challenge that this type of data presents for all fields of data science, not only humanitarian and human rights applications of ICTs and the data derived from them.

The following definition should be seen as the first provisional step towards an initial pedagogy exploring DII, its uses, ethical dimensions, and the challenges it presents to practitioners in this field and many others. For the purposes of this chapter, DII is defined as follows:

> Demographically Identifiable Information, or DII, is defined as either individual and/or aggregated data points that allow inferences to be drawn that enable the classification, identification, and/or tracking of both named and/or unnamed individuals, groups of individuals, and/or multiple groups of individuals according to ethnicity, economic class, religion, gender, age, health condition, location, occupation, and/or other demographically defining factors.
>
> DII can include, though is not limited to, personal identifiable information (PII), online data, geographic and geospatial data, environmental data, survey data, census data, and/or any other data set that can – either in isolation or in combination – enable the classification, identification, and/or tracking of a specific demographic categorization constructed by those collecting, aggregating, and/or cross-corroborating the data.

The importance of DII in civil society applications of ICTs and the data derived from them cannot be overstated. It may be argued that most, if not all civil society applications of ICTs and the data derived from them fundamentally aim to collect, analyze, and create actionable products either initially based upon and/or seeking to result in DII. DII can be seen as, at first glance, ethically neutral by itself in many cases, without a seemingly obvious ethical imperative for a practitioner to immediately act upon.

The 2013 Red Cross Professional Standards for Protection Work, comparing the risks of aggregated data to sensitive individually identifiable data, seems to underplay the risks of these aggregated data sets, stating:

> Protection actors working with aggregated information, such as trend analysis, do not face the same challenges as the information they handle is less sensitive. They may feel less concerned by the standards and guidelines of this chapter. They should nevertheless be aware of the constraints of managing data on individuals and events, in order to understand how the information they are handling has been obtained. (ICRC 2013)

The more seemingly subtle ethical implications of DII are in stark contrast to many common types of PII encountered in the civil society context, such as raw, de-identified individual health records or refugee registration documents. DII's ethical implications largely results situationally from when, how, why, and from what combinations of initial sources it is derived and applied, rather than the more easily ethically categorized data that comprises PII.

In other words, DII can result from the transformation of seemingly disparate, unrelated data sets into a an amalgamated data product that can be easily "weaponized" into a means for doing harm. The potential harm of DII is often most apparent, if not entirely, to the perpetrator of potential harm, rather than to the holder of one or all of the pieces of a potentially actionable mosaic of DII.

Whereas PII's potential harm comes from when it is leaked or breached, DII's harm, and thus its ethical implications, often emanates from simply whether the possibility exists that it can be even created. This reality makes the overall ethical imperative to understand, manage, and protect potential sources of DII as important, if not more so in some cases, than those commensurate with holding only one source of PII.

4.8 DII: A Hypothetical Example

The following hypothetical example seeks to illustrate the basic chain by which DII can be created and acted upon by potential perpetrators of abuses against a vulnerable population. In this scenario, an NGO managing several displaced persons camps in country X has allowed a UN agency to publish a map showing the camps with the largest population influxes of displaced people in recent months. Sensitive infrastructure, in particular a protection center for demobilized child soldiers, have been excluded from the maps to protect vulnerable demographics residing in the camps.

Meanwhile, an agency working to assist the demobilized child soldiers at the protection center has published an online blog stating that it is providing services to these children at an unnamed camp that has experienced the largest influx of displaced people. A non-state armed actor seeking to reclaim child soldiers that had previously fought in its group cross-corroborates the de-identified map with the detail about the displaced person influx at the camp in the de-identified blog story to locate where the former child soldiers are living, enabling them to attack the camp and abduct the children.

In this hypothetical scenario, individual informed consent was neither required nor violated; PII information was not collected to create either data stream; and a DII product that gave an armed actor otherwise unavailable actionable information was created from seemingly benign, separated, de-identified, and open source information. Though none of the agencies who produced the individual products acted with maleficence (the intent to do harm), they could be accused of failing to engage in non-maleficence by not anticipating how the two data points, when fused together, created a targeted DII product that put the children at risk.

This example should not be taken as encouraging a prohibition on either public information sharing by civil society actors that promotes their operational activities, nor discouraging the creation of internal information products in the service of the duty towards beneficence that employ DII. It is simply to demonstrate that the one of the primary, emerging ethical challenges the field must face in order to act with non-maleficence requires core competencies, tested methodologies, and ethical protocols that do not currently exist.

Additionally, this example also shows that traditional conceptions of PII and individual consent are insufficient by themselves to address the rapidly evolving matrix of uniquely twenty-first century data-based threats. The bedrock principles of protecting PII and seeking of individual informed consent, when applicable, should continue to be anchoring components of any ethical regime for civil society organizations. However, these two approaches cannot simply be retrofitted with new norms, protocols, and training to meet these challenges.

Instead, these core concepts must be supplemented by new science and new ethical principles specific to these new threats and challenges, rather than simply either augmenting or discarding what currently exists. A concerted, collaborative, and clear-eyed effort by researchers, ethicists and practitioners working together is required to equip those on the front lines of data driven response with the capacities necessary for ethical action. Only through an honest and transparent sharing of past incidents and current challenges can an actionable theory of DII be developed to inform new frameworks for guiding organizations in contexts when traditional informed individual consent cannot be obtained or does not apply.

While many of these threats and challenges cannot be prevented or eliminated, they can be better mitigated if the field reaches consensus based on available evidence that these new dynamics exist and are not going away. The current approach can be summarized as attempting to "do no harm" without actually knowing what the full extent of the harm might be, nor how that harm is both created and manifested.

"First, doing no harm" without first "knowing the harm" is thus impossible. The ethical complexities presented by DII are only compounded by the fact that this type of data is often generated in contexts where traditional individual informed consent is not only impossible, but is actually the motivating factor for choosing remote data collection platforms with the goal of generating PII.

4.9 Multi-stream DII Collection and the "Consent Paradox"

An illustration of a scenario where multi-stream, DII data was the goal of a civil society organization's data collection is the pilot phase of the Satellite Sentinel Project (SSP) from 2010 to 2012.[4] SSP integrated ground reporting with the analysis of high resolution satellite imagery to create public reporting about alleged attacks on civilians and apparent threats to vulnerable populations in Sudan and South Sudan.

The multi-stream approach often specifically focuses on deriving otherwise unavailable data about demographic groups, as was the case with SSP. Information gained through the fusion of multiple data streams by SSP included the locations, movement patterns, size, and apparent status of both civilian populations and armed actors largely without using what would traditionally be seen as PII data.

SSP's approach highlights the issue at the core of the "Consent Paradox": Civil society organizations deploying large scale ICT interventions for the collection of digital data often do so precisely *because* the type of ground access necessary for individually consented data capture is impossible. My colleagues and I write of the operational context in which SSP was deployed as the reason necessitating the use of large scale, multi-stream data collection focused on particular ethnic groups, stating:

> At the time SSP was launched, approximately two weeks before the January 2011 referendum deciding southern Sudan's secession from Sudan, credible data about events on-the-ground were scant. Violence was escalating. Specific ethnic groups in Abyei, Blue Nile, and South Kordofan were seen by analysts as potential targets for atrocities by the government of Sudan. The information available about the events in these areas was often second-hand and largely impossible to confirm. The international community had minimal capacities for collecting impartial information and freely assisting civilians inside critical areas of Sudan due to restrictions on their freedom of movement.

The "Consent Paradox" can thus be defined as when organizations, who likely seek to live the principle of humanity through trying to obtain informed consent whenever possible, are forced to impossibly balance that expectation with the operational requirements of working in inherently non-permissive environments. Consequentially, organizations may be increasingly caught between either abiding by established but outdated ethical norms with no clear alternative approach identified versus a perceived life saving opportunity for potentially increased situational awareness and operational impact.

The Consent Paradox will likely persist as long as there is no alternate ethical paradigm for attempting to achieve non-maleficence and guide the delivery of beneficence in these settings and operational contexts other than informed individual consent. This reality, while understandable, is not ethically tenable.

At present, humanitarian and human rights agencies are being forced to contend with the Consent Paradox as they respond to an unprecedented number of "Level 3"

---

[4]Note: The author served as the founding director of operations for SSP from December, 2010 until the summer of 2012.

humanitarian crises as of 2015 – all of them conflict related. Protracted responses to emergencies in Syria, South Sudan, Yemen, and Iraq likely compel organizations to engage in data-based action in highly dangerous twenty-first century environments with largely untested technologies and methodologies far outside the bounds of the available twentieth century ethical guidance. Meanwhile, the consequences of this inherently experimental data action on the affected communities these groups seek to serve are largely unknown and often unmeasurable.

Thus, organizations seeking to embody the ethic of non-maleficence primarily through the protection of PII and a reliance on individual informed consent models are pursuing an ethical paradigm that is, in an increasing number of cases, anachronistic. Evolving collection approaches, uses, and operational contexts have rendered a PII and individual consent-focused ethics alone insufficient in an increasingly evolving and complex networked world – a world that is quickly superseding the traditional normative frameworks available to these actors.

4.10 The Population Protection Imperative: Towards a New Framework for Civil Society's Use of Data

The existing PII and individual informed consent paradigm places the ethical focus of organizations overwhelmingly on how the data individual organizations use was collected, analyzed and stored by the organization seeking to employ it for a specific operational or programmatic purpose. The complex and currently poorly understood challenges of DII, as well as the potential demographic and community-based impacts of data threats, are not countenanced by current approaches.

If this crucial blind spot is not directly addressed by the field at large, then organizations that are continuing to deploy data-based approaches to support their work will be likely creating and magnifying threats to the vulnerable populations they seek to serve without the means to identify and mitigate these threats. This paradigm is thus fundamentally unethical – even if these deployments are compliant with traditional PII and individual consent standards.

It also may violate civil society's ethical duty to act with non-maleficence, likely leading to further, inevitable harm to already vulnerable people. Additionally, it limits the potential beneficence of these activities because an understanding of both potential harms and benefits is not currently preceding the design of data-based interventions in a standardized way.

A profound and difficult pivot is now required of civil society organizations to deploy data more ethically in the face of the complexities of an increasingly DII-based data ecosystem. Rather than the current approach, which is a "Data Protection Imperative" predominantly focused on safeguarding PII data and the individuals it is derived from, a "Population Protection Imperative" must instead be articulated and implemented in a commonly routinized way. Developing this new approach is incumbent on civil society organizations if these new and increasingly prevalent modalities for data collection and use are to begin to be considered ethically applied.

This new approach must place the onus on organizations to first consider and respond to the external context of factors that can make data harmful, as opposed to simply placing the emphasis on whether an organization can be internally responsible for primarily managing only the data it collects and uses. Fundamental components of operationally realizing a Population Protection Imperative as part of civil society's current use of data may include the following:

- *First, know the harm before seeking to do no harm*: An organization's capability and capacity to identify, detect, and reasonably mitigate what potential vectors for harm exist both in its internal collection and use of data, as well as the broader external context in which it is acting, must become an organization's first ethical responsibility. To achieve this paradigm, a new science of DII is required to equip organizations with methodologies and an evidence-base for determining the potential harm of DII as a core aspect of acting with non-maleficence.
- *"Touch me not!"*: If an organization determines it is unable to reasonably know the potential harm to a basic level of certainty, then it is unable to claim to do no harm, and thus must cease its project immediately. This concept in ethics is referred to as *Noli Me Tangere*, or "Touch me not!". There must be a willingness and a capability to suspend activities when the potential ethical consequences are deemed to be unknowable to a degree that precludes the development of basic management and mitigation strategies. In many cases, current practice does not provide clear guidance to organizations about how and when to address these increasingly commonplace scenarios. While the humanitarian principle of humanity may seem to conflict with a "Touch me not!" ethic, continuing the current approach encourages the type of "disaster experimentation" that McDonald identifies in the context of the ebola crisis. "Saying no" to deployments where the harm cannot be known or mitigated should be seen as a core part of living the humanity principle and respecting the dignity of affected populations.
- *Inter-organizational coordination versus internal curation*: The focus must shift from organizations prioritizing their ability to internally curate data that they collect to a new and fundamentally different focus on being able to coordinate why and how data will be collected amongst a diverse group of organizations present in a specific operational context. The threats inherent in DII data make collaboration essential to addressing them because, unlike PII data, the aggregate combination of data being collected by all organizations is where the threat originates and resides. While seemingly obvious, this collaborative approach to the initial development of data management strategies, which goes far beyond "information sharing" alone, is a profound challenge given the often competitive nature of civil society data deployments. However, it is required if a new science and a new ethics of data use appropriate for the emerging dynamics of the DII-based data ecosystem in which organizations are operating is to be realized.
- *Demographic threat triage and transparent after-action:* Organizations must begin to develop and implement the capacity to proactively triage what data (and in what combinations) present which identifiable types of potential threats to specific demographics. It is from this methodology that pre-deployment review mechanisms for funders and data actors can be developed. This front-end process

also requires a corresponding back-end mechanism of after-action review that is transparently performed and widely shared within the community. The current dearth of available evidence of threats and their potential impact can only be addressed through this twin approach of pre-deployment triage and after-action review if the non-maleficent use of data is to be achieved.

- *Data preparedness must precede data collection:* For civil society organizations to achieve their duty for beneficence, a scalable and widely implemented concept of data preparedness (e.g. a methodology for ascertaining what data is required to have what operational impact in certain contexts) is urgently required. In many cases, civil society organizations collect data based on what is possible to collect, rather than collecting data because of an evidence-based assessment of what is needed to achieve what impacts. The duty of beneficence requires organizations to understand what data can provide what benefits in specific scenarios prior to the decision to even deploy data collection modalities.

In conclusion, the current ethical state of the art of civil society's use of data that inherently affects demographic groups, rather than individuals alone, is insufficient to meet the poorly understood and rapidly evolving nature of potential threats and harms this work may create and cause. Civil society must begin to recognize that the challenges of protecting PII and obtaining individual informed consent are quickly being superseded, though not replaced, by the emergence of DII and the demographically manifested implications of large scale digital data collection.

While the twentieth century ethical architecture and the responsibilities it seeks to fulfill must remain, a new twenty-first century ethical architecture must be urgently developed to supplement the previously extant normative framework to a degree that addresses these complex and evolving challenges. This new framework should include the development of an evidence-based for understanding and managing DII and the situations where individual consent may neither apply or can be obtained. Also, these new frameworks must be based on a collaborative inter-organizational approach that recognizes the shared nature of these threats and the common responsibly across all groups to address them together.

There are no easy answers to these challenges. However, continuing to apply outdated ethical constructs to modalities of data collection and manifestations of data threats that were not countenanced in the pre-digital age is not an option. Opting out of the "brave new world" is not a viable or responsible choice for organizations either. While the way forward is not clear, the responsibility of civil society to innovate its capacity for ethical action equal to its new technological capabilities has never been more clear.

Bibliography

Al Achkar, Z., B. Card and I. Baker. 2013. *Sharing space: Adapting military approaches to geospatial analysis for humanitarian response and the documentation of human rights abuses.* Retrieved from http://www.hpcrresearch.org/sites/default/files/publications/2%20Sharing%20Space_HHI_Final_a%20copy_0.pdf

Chamales, G., and R. Baker. 2011. *Securing crisis maps in conflict zones*. Global Humanitarian Technology Conference (GHTC), 2011, 426–430. IEEE.

Code of Conduct. 2014, March 20. *Standby task force*. Retrieved from http://blog.standbytaskforce.com/our-model/code-of-conduct/.

De Montjoye, Y., J. Kendall and C. Kerry. 2014. Enabling humanitarian use of mobile phone data. Series: Issues in technology innovation. *The Brookings Institution*. Retrieved from: http://www.brookings.edu/research/papers/2014/11/12-enabling-humanitarian-use-mobilephone-data.

Harvard Humanitarian Initiative. 2012, July. *Making the world a witness: Report on the pilot phase*. Retrieved from http://hhi.harvard.edu/sites/default/files/publications/making-the-world-a-witness.pdf.

ICRC. 2013. *Professional standards for protection work: Carried out by humanitarian and human rights actors in armed conflict and other situations of violence*. Geneva: ICRC.

McDonald, S. 2016. *Ebola: A big data disaster*. The Centre for Internet and Society. Retrieved from http://cis-india.org/papers/ebola-a-big-data-disaster.

Oxfam. 2015, August 27. *Responsible program data policy*. Retrieved from http://policy-practice.oxfam.org.uk/publications/oxfam-responsible-program-data-policy-575950?intcmp=RM_ResponsibleDataPolicy.

Pantilat, S. (n.d). *Beneficence vs. Nonmaleficence*. Retrieved from http://missinglink.ucsf.edu/lm/ethics/ContentPages/fast_fact_bene_nonmal.htm.

Raymond, N.A., and B. Card. 2015. *Applying humanitarian principles to current uses of information communication technologies: Gaps in doctrine and challenges to practice* [White Paper]. Retrieved from http://hhi.harvard.edu/sites/default/files/publications/signal_program_humanitarian_principles_white_paper.pdf

Raymond, N., C. Howarth and J. Hutson. 2012, February 6. *Crisis mapping needs an ethical compass*. Retrieved from http://globalbrief.ca/blog/2012/02/06/crisis-mapping-needs-an-ethical-compass/.

Raymond, N., B. Davies, B. Card, I. Baker, and Z. Al Achkar. 2013. While we watched: Assessing the impact of the satellite sentinel project. *Georgetown Journal of International Affairs* 14(2): 185–191.

Raymond, N.A., B. Card and Z. Al Achkar. 2015. *What is 'Humanitarian communication'?: Towards standard definitions and protections for the humanitarian use of ICTs*. Retrieved from https://www.eisf.eu/wp-content/uploads/2015/10/2041-EISF-2015-What-is-humanitarian-communication.pdf.

Rendtorff, J. 2009. *Responsibility, ethics, and legitimacy of corporations*. Frederiksberg: Copenhagen Business School Press.

Smith, C.M. 2005. Origin and uses of primum non nocere — Above all, do no harm! *The Journal of Clinical Pharmacology* 45(4): 371–377. doi:10.1177/0091270004273680.

Humanitarian Accountability Partnership. 2010. The 2010 HAP Standard in Accountability and Quality Management. 2010. Retrieved from http://www.hapinternational.org/pool/files/2010-hap-standard-in-accountability.pdf.

U.S. Department of Health and Human Services. n.d. *What is informed consent and when, why, and how must it be obtained?* http://www.hhs.gov/ohrp/policy/faq/informed-consent/what-is-informed-consent.html.

U.S. Department of Labor. n.d. *Doing business with the department of labor – Guidance on the protection of personal identifiable information*. Retrieved from http://www.dol.gov/dol/ppii.htm.

UAViators Code and Guidelines. n.d. Retrieved from https://docs.google.com/document/d/1Uez75_qmIVMxY35OzqMd_HPzSf-Ey43lJ_mye-kEEpQ/edit.

Chapter 5
Group Privacy: A Defence and an Interpretation

Luciano Floridi

Abstract In this chapter I identify three problems affecting the plausibility of group privacy and argue in favour of their resolution. The first problem concerns the nature of the groups in question. I shall argue that groups are neither discovered nor invented, but designed by the level of abstraction (LoA) at which a specific analysis of a social system is developed. Their design is therefore justified insofar as the purpose, guiding the choice of the LoA, is justified. This should remove the objection that groups cannot have a right to privacy because groups are mere artefacts (there are no groups, only individuals) or that, even if there are groups, it is too difficult to deal with them. The second problem concerns the possibility of attributing rights to groups. I shall argue that the same logic of attribution of a right to individuals may be used to attribute a right to a group, provided one modifies the LoA and now treats the whole group itself as an individual. This should remove the objection that, even if groups exist and are manageable, they cannot be treated as holders of rights. The third problem concerns the possibility of attributing a right to privacy to groups. I shall argue that sometimes it is the group and only the group, not its members, that is correctly identified as the correct holder of a right to privacy. This should remove the objection that privacy, as a group right, is a right held not by a group as a group but rather by the group's members severally. The solutions of the three problems supports the thesis that an interpretation of privacy in terms of a protection of the information that constitutes an individual—both in terms of a single person and in terms of a group—is better suited than other interpretations to make sense of group privacy.

Keywords General Data Protection Regulation (GDPR) • Data protection • data ethics • Individual privacy • Group privacy

---

L. Floridi (✉)
Oxford Internet Institute, University of Oxford, 1 St Giles, OX1 3JS, Oxford, UK
e-mail: luciano.floridi@oii.ox.ac.uk

L. Taylor et al. (eds.), *Group Privacy*, Philosophical Studies Series 126,
DOI 10.1007/978-3-319-46608-8_5

5.1 Introduction

The debate on Big Data (including Open Data) and Data Protection focuses on individual privacy. How can the latter be protected while taking advantage of the enormous potentialities offered by ever-larger data sets and ever-smarter algorithms and applications? The tension is sometimes presented as being asymmetric: between the *ethics* of privacy and the *politics* of security. In fact, it is ultimately ethical. Two moral duties need to be reconciled proactively: fostering human rights and improving human welfare. The tension is obvious if one considers medical contexts and biomedical big data, for example, where protection of patients' records and the cure or prevention of diseases need to go hand in hand.[1]

Currently, the balance between these two moral duties is implicitly understood within a classic ontological framework. The beneficiaries of the exercise of the two moral duties are the individual person vs. the whole society to which the individual belongs. At first sight, this may seem unproblematic. We work on the assumption that these are the only two 'weights' on the two sides of the scale. Such a framework is not mistaken, but it is dangerously reductive, and it should be expanded urgently. For there is a third 'weight' that must be taken into account by data protection: that of groups and their privacy.

The chapters in this volume provide a detailed analysis of the possibility of attributing a right to privacy to groups and sophisticated analyses of the scholarship behind the debate on group privacy, especially in modern legislation. In this contribution, I shall assume that it is *prima facie* plausible that groups may indeed enjoy such a right. However, there are at least three problems that may undermine such plausibility. I shall address them in the following pages with the hope that their solutions will facilitate the development of our ideas on group privacy.

The first problem, to be discussed in Sect. 5.2, concerns the nature of the groups in question. I shall argue that groups are neither *discovered* nor *invented*, but *designed* by the level of abstraction (LoA) at which a specific analysis of a social system is developed. Their design is therefore justified insofar as the purpose, guiding the choice of the LoA, is justified. This should remove the objection that groups cannot have a right to privacy because groups are mere artefacts (there are no groups, only individual persons to which groups are ultimately reducible) or that, even if there are groups, it is too difficult to deal with them.

In Sect. 5.3, I shall address the next problem: assuming that there are groups and that they can be successfully managed, in what way can they be attributed rights? I shall argue that the same logic of attribution of a right to individual persons may be used to attribute a right to a group, provided one modifies the LoA and now treats the whole group as an individual in itself. I shall further argue that attributing a right to a person *or* to that person's group need not be incompatible alternatives, that is, the 'or' may be sometimes read as inclusive (as a logical 'and/or' or Latin *vel*, not neces-

[1] (Howe et al. 2008) and (Groves et al. 2013), for a review see (Mittelstadt and Floridi 2016). Most recent analyses of ethical problems in biomedical big data are provided in (Mittelstadt and Floridi forthcoming-b).

sarily always as an *aut aut*). This should remove the objection that, even if groups exist and are manageable, they cannot and should not be treated as holders of rights.

In Sect. 5.4, I shall then show in what sense groups may enjoy a right to privacy as groups. This should remove the objection that privacy, as a group right, is a right held not by a group as a group but rather by the group's members severally. Sometimes it is the group and only the group, not its members, that is correctly identified as the correct holder of a right to privacy. The analogy here is with the right of self-determination, which is held by a nation as a whole, not merely by its members severally.

The solutions of the three problems listed above lead to a final set of considerations, in Sect. 5.5, about the nature of privacy that may be enjoyed by a group. There I shall argue that an interpretation of privacy in terms of a protection of the information that constitutes an individual—both in terms of a single person and in terms of a group—is better suited than other current interpretations to make sense of group privacy.

To conclude, I shall argue that there are groups, designed by our ways of modelling interactions between agents and patients (senders and receivers of actions); that they can be and are manageable as holders of rights; and, in particular, that groups can be the primary holders of a right to privacy when this is constitutive of their identities. If I am correct, there is plenty of work for legislators to do. Let us see whether I am.

5.2 How Can There Be Groups?

The debate about the nature of groups in philosophy of law and social science is strictly related to two other debates. One, in analytic philosophy (Beebee and Sabbarton-Leary 2010; Campbell et al. 2011), concerns natural kinds and whether there are 'natural'—as opposed to only arbitrary—ways of grouping objects, events, or beings on the basis of some shared, intrinsic properties (essentialism), such as chemical composition in the case of all objects made of gold (where gold is the natural kind). The other debate, in philosophy of biology (Panchen 1992; Laporte 2004; Richards 2010; Oderberg 2013), concerns the nature of species, and more generally biological taxonomies, and addresses the question whether one or more criteria (such as reproductive isolation, or what it looks like, or indeed is, genome) may be sufficient to categorise species, or decide whether an organism belongs to one species or another.

The similarities between the three debates are due to the fact that they are particular versions of the more fundamental and long-term debate between nominalism (there are only individuals, or tokens) and realism (there are also universals, or types). The nominalists and the realists tend to agree on the existence of individuals. They are both happy with Alice, her golden ring, and her puppy. They disagree on the existence of groups (Alice's family), natural kinds (golden objects), and species (Canis familiaris), and, in some cases, on the order of ontological priority (in various forms of Platonism universals not only exist but also precede, in terms of logical order, their instantiations). In short, they disagree on whether groups, natural kinds, and species may be only subjective and observer-dependent, or also objective and observer-independent.

Such ontological disagreement about what there is in reality and how it is organised in itself is possible because it presupposes a common epistemological framework, which enables the nominalist and the realist to avoid arguing at cross purposes. This is the view that knowledge can provide direct access to the intrinsic nature of its referents, i.e. what there is (or isn't) in the world in itself; the *noumenon*, to use a Kantian terminology. Interestingly, the further we move away from natural sciences and the closer we get to social or engineering ones, the easier it is to see this as a mistaken assumption, which leads to a false dichotomy. According to the nominalist, social groups (to restrict now the issue to our current concerns) are *invented*. According to the realist, they are *discovered*. The truth is that they are *designed*, that is, they are the outcome of the coming together of the world and the mind. To be more precise, they result from the choices we make of the observables we wish to focus on, for specific purposes, and from the constraining affordances (data) provided by the systems we are analysing. Thus, the position I wish to defend about the ontology of social groups is anti-essentialist but not anti-realist.[2] Let me illustrate it with an analogy.

Let us call a set of observables a level of abstraction (LoA). There is a LoA at which there are only individual buildings, and Alice's new flat and Bob's Victorian semi-detached house cannot possibly form a group. The two buildings may be regulated by very different kinds of legislation, provide different affordances, appeal to different home buyers, and so forth. They are so different from each other that they could never form a group. But then there is also a LoA at which both are two-bedroom accommodations in Oxford, for example, subject to the same local council taxation, perhaps rented from the same owner, and so forth. They are obviously part of a group. Asking whether a set of entities does or does not form a group independently of why one is asking the question in the first place, that is, independently of any interest in which features of the objects should count (e.g. the number of bedrooms for taxation purposes) is like asking the absolute price of a car without accepting any currency as a means to convey it.

There are of course groups that seem to us more natural. Yet the naturalness of a grouping is just a function of the intuitiveness of a LoA, that is, it is epistemological, not ontological. Referring to salad, tomatoes and potatoes as a group called food seems something as observer-independent and objective as possible, but this is only because we assume our own interests as organisms and eaters as the natural, intuitive, and relevant LoA. To a tiger, they would all look as unrelated and as eatable as grass and leaves to us. Accepting that our knowledge of the world is obtained through different LoAs is not to say that anything goes, and that the only alternative to nominalism and realism is some kind of untenable relativism. It is to say that absolute questions asked in a logical space lacking any references (LoA) and orientation (interest, purpose) are an absolute mess, and that *relationalism* (or *liminalism*, if you prefer a fancier word) is a better alternative. Using the previous example,

[2] For a similar position in philosophy of biology see (Khalidi 2013).

asking whether something is food means adopting the right LoA at which it makes sense to ask whether a specific substance can be a nutrient for a specific organism. Food is a relational (not a relative) concept: it takes a LoA with two relata to define it, yet not every LoA is correct and some LoAs will be more correct than others.

All this means that we cannot be naïvely nominalist or realist about our ontology, especially when it comes to complex objects such as social groups. Imagine reality in itself as a sender of messages. Reality, understood as the Big Radio, broadcasts a very wide spectrum of signals. We, humans, are able to receive some of them directly, some others indirectly. For example, the visible spectrum is the portion of the electromagnetic spectrum that is detectable by the human eye and this is our most fundamental LoA when it comes to visual perception; we can see invisible radiant energy (for example, infrared, electromagnetic radiation with longer wavelengths than those of visible light) through technological mediations. Out of all those signals, we make sense of the sender itself. It would be utterly naïve to think that the signals are a description of the sender, yet this does not mean that they are any less real. We only have to admit that the Big Radio is not sending selfies. With two other, different analogies, we cook with some ingredients (data from the world) but the dish we obtain (information) is not a copy of the ingredients. Or, we build with some materials (data in the world), but the house we obtain (information) is not a copy of the bricks we used. Human knowledge works in this *constructionist* (not *constructivist*, mind) way, it is not *mimetic*, it is *poietic*. Some parts of this *poiesis* are heavily constrained by the signals we receive. In the long run, we ask more questions to get more data, as Francis Bacon already suggested. We manipulate the data to see what further data can be obtained, and all this leads to scientific theories, which are our best ways of making sense of the constraining affordances (my preferred definition of data) provided by the realities we are studying. Some other parts of this interpretation are more flexible and malleable, i.e. the constraining affordances provide much more latitude, and well-informed, rational disagreement is more difficult to resolve (think of economic policies during a financial crisis). There is nothing relativistic or anti-realist in this, in the same sense in which there is nothing relativistic or anti-realist in the dish we cooked or the house we built. Humanity has taken advantage of the signals sent by the Big Radio increasingly well and this is why our knowledge works so successfully. The fact that we find some grouping very intuitive is part of such a successful story. But we do not need to embrace any naïve essentialism, or representational theory of knowledge, or a correspondentist theory of truth to make sense of groups. We should think about our knowledge of the world not in terms of painting it but in terms of engineering a model of it. Grouping is part of the successful strategy through which we make sense of reality.

What follows from the previous outline is that social groups should neither be conceived as mere conventions or artefacts (invented) nor assumed to exist before the interest in identifying them is specified (discovered). They are more or less correctly and successfully designed by our epistemological interests and practices *together* with the ontological constraining affordances provided by the world.

Let us now return to the nature of social groups. Any social system of n individuals can be organised into 2^n groups. For example, Alice, Bob and Carol would give rise to the following eight groups (subsets): {}, {Alice}, {Bob}, {Carol}, {Alice, Bob}, {Alice, Carol}, {Bob, Carol}, {Alice, Bob, Carol}. It is obvious that the power set of a set (the group containing all possible groupings) soon becomes unmanageable. At the same time, privileging only some groups as 'real' may seem to be arbitrary. Why should {Alice, Carol} count, but not {Bob, Carol}? Because both Alice and Carol are female? But what if the criterion is having a rare disease, which Bob and Carol share, but not Alice? Clearly what matters is the LoA (in our example gender or health) at which the data we have (in our example, Alice's, Bob's, and Carol's)—the constraining affordances—are transformed (modelled) into information that ends up generating a group. The logical order is therefore: purpose (why grouping individuals in this way), LoA (how grouping individuals in this way), result (the obtained group). With an elementary example,[3] in a legal class action first comes the interest in dealing with a specific issue. This sets the observables (the LoA), e.g. some Electrolux dryers are alleged to "contain defects that can cause them to catch fire due to lint buildup". Given this LoA, one can then identify the group, that is, who is eligible "if you purchased certain freestanding clothes dryers between Jan. 1, 2002 and Dec. 31, 2011, you could be eligible for benefits from the Electrolux class action settlement". The LoA designs the group of eligible people. Asking whether the group is discovered objectively or invented subjectively *before* the interest and LoA are specified is not even incorrect, it is just missing the point entirely. Of course, some social groups simply self-determine their own nature, by adopting the purpose and LoA at which they wish to be identified.

All this is particularly relevant in the case of group privacy because it would be a mistake to think that first one has to establish the existence of a group, then the presence of a group's right to privacy, and then the potential infringement of that group's privacy through some Information and Communication Technologies (ICTs) application. If this were the case, we would be facing an intractable problem, because the identification of groups *a priori*, independently of the identification first of any interest or purpose (and hence LoA) that determines the grouping, is open to endless debate. Luckily, the process in practice is rather the opposite. First comes the interest (usually, but not necessarily pursued through the application of a technology) in clustering people in some groups. For example, a retailer may be interested in reaching all pregnant women in Oxford in order to advertise some products. This group may or may not overlap with other, pre-existing, intuitive groups, yet this does not matter (although this can be confusing when approaching the issue from a nominalist vs. realist perspective), even when the interested practices in question may be self-reflective, i.e. even when individuals may wish to identify themselves as members of a group, for this too is an epistemological choice. (note that the mistake here would be to attempt to identify all possible social groups in Oxford and then check whether their rights have been infringed, an impossible task). Then comes the potential breach of the privacy of such a group as a group ('as a group' is an assumption that still needs to be defended below, bear with me). Note that what constitutes the group is also what makes group privacy possible. And finally comes the right of the group to

[3] Seehttp://topclassactions.com/lawsuit-settlements/open-lawsuit-settlements/30306-electrolux-dryer-class-action-settlement/

see the situation redressed. In short, there is no nominalist objection to group privacy because it is the very same interested practices determining the grouping of people that also delineate the resulting groups as potential holders of a right to privacy, which then the group can exercise. Profiling is not a *descriptive* practice, it is a *designing* one, and it comes with the consequence of creating the condition of possibility of the profiled individuals, now constituted as a group by the very act of profiling, to act as a group in order to claim respect for its own privacy. Of course the grouped (profiled) individuals may not know that they have been profiled, e.g. by automatic algorithms, and may never discover that they have been treated as a group. This is not the point. What I am arguing is that if they end up being profiled and this profiling becomes explicit, what gives the group the initial possibility of reacting to it is the "interested" practice of profiling it in the first place, not some pre-existing ontological status of the group as a group, that would allegedly predate the profiling. With an analogy, the slice may not know that is has been severed from the rest of the cake, but if it realises that it has been it also realises that it was the severing it from the cake that gave rise to its identity, which did not precede the severing process itself. With one more analogy, grouping cuts both sides of the same piece of paper, the social (who is and is not in a group) and the ethical (which group has a right to privacy); you cannot have one without the other. All this explains why profiled individuals often object not so much to the treatment of themselves as members of a group but to the very profiling in the first place (it is not being a slice the problem, the problem is being severed from the cake in the first place).

The next question then becomes: if groups are constituted by the interested practices of grouping, for a purpose, and at a particular LoA, in what sense, if any, can they have a right?

5.3 How Can a Group Have Rights?

Groups are the social, qualitatively richer instance of mathematical sets. This is useful, because, by looking at sets, it is much easier to clarify in what sense a group and its members may or may not share the same property, including a particular right. Let me explain.

Imagine a small departmental library. We need to move it from one building to another. We decide to move first all books with authors from A to D. Clearly the pile of books does not share that property, that is, it would be meaningless to ask whether the pile has an author. Next, suppose we are concerned about the fact that each of our books is inflammable. The concern remains once we realise that the pile inherits the same property. Third, we try to lift the pile and notice that it has now acquired a property that none of the books has: it is too heavy to be moved by a single person, despite the fact that each book in it is reasonably small and light. With a sigh, we finally wish books could fly from one building to another, but they do not, and neither do piles of them. This example illustrates the four possible cases in which sets and their members may or may

	has the property *F*			
	1	2	3	4
Members	Yes	Yes	No	No
Set	No	Yes	Yes	No
Example	Author	Inflammable	Heavy	Flies

Fig. 5.1 The relations of commonality of properties between sets and their members

not share a property (see Fig. 5.1). I introduced them in order of importance. The first case generates a common fallacy. The last case is not relevant to our discussion.

The debate on whether groups (sets) may have rights (the property *F*) can be clarified by using the four columns in Fig. 5.1.[4] Sceptics subscribe to position 1: rights are properties that qualify only members of a group, not a group; speaking of a group right makes no sense and it is based on a fallacy. Moderate supporters of group rights tend to sit in the middle, subscribing to position 2: a group has rights, but only because each individual person constituting it has such rights. Finally, strong supporters of the idea of group rights subscribe to position 3: there are some kinds of rights that belong only to a group as a group, not to a group insofar as it is constituted by individual persons who enjoy those rights. In this case, it is important to understand that the group itself acts as an individual, to which a right is attributed. This is the case with political rights, as we have already seen: it is a shift in the LoA that allows one to consider a whole nation as having a right to self-determination as an individual agent. The point is important not only for the sake of clarity, but also because we saw that determining the LoA is what makes talking about groups ontologically unproblematic. By grouping people according to specific criteria we create an individual (the group), which can both be targeted and claim to have rights as a group.

The debate between the sceptical, the moderate and the strong position about group rights leads us to the last problem I wish to address here: how a group can have a right to privacy.

5.4 How Can a Group Have a Right to Privacy?

One problem with privacy is that it is unclear whether, if it applies to groups, it may apply sometimes in the moderate and sometimes in the strong sense. Consider the following two cases.

[4] For the sake of simplicity in what follows I shall assume that if members of a set and the set have the same property *F* this is because the set inherits *F* from its members. This is not necessarily the case and things become more complicated if we include the case in which both members and their set may have the same property *F* but for different reasons, that is, if the relation between the *F* of the members and the *F* of their set is not one of inheritance but of repeated occurrence. For example, the set of all books without an author is also without an author, but not because of them, but because authorship does not qualify sets of books, only books.

A new California Privacy Law for Minors took effect as of January 1, 2015.[5] Entitled "Privacy Rights for California Minors in the Digital World", it gives minors the right to delete content that they posted to a website, social media profile, or online service while under the age of 18. It also includes restrictions on marketing or advertising some specified products and services to minors. This law seems a case of moderate group privacy. It is phrased in terms of protection of the individual person (the term "minor" is used, in line with Privacy Law, to mean natural person individual under the age of 18 who resides in California) and it seems obvious from the text that any reference to minors as a group (the "General Audience Property(ies)") is only a shortcut for a reference to each of its members. Minors have a right to see their personal information online erased only because each minor does. Talking of group privacy in this case is merely convenient but does not seem to add anything to our understanding of the phenomenon.

Consider next the case in which the close friends and relatives (the group) of a deceased person decide to hold a private funeral. Attendance is by invitation only, but this is not meant to make the funeral 'exclusive'. The desired privacy may be due to a need for intimacy, for respectful quietness, to protect grieving and reflection, or perhaps because of cultural or religious customs. Whatever the reasons, in this case it seems very counterintuitive to argue that each member of the group (each close friend or relative of the deceased) has a right to a private funeral, or that the privacy demanded is just the collection of all individual privacies. It seems more reasonable to admit that we are in the presence of a strong, social sense of group privacy. It is the whole group as a group that has a right to that specific kind of privacy.

If privacy applies to groups only in the moderate sense seen above (recall also the analogy with the pile of books, which is inflammable just because each book in it is), then there is interest in exploring its consequences, but not its nature. For if groups have a right to privacy only insofar as their members do, then all that can be said about moderate group privacy in terms of theory can also be said by reference to personal privacy (there is nothing special in group privacy over and above all the personal privacies of the group members), yet this very reducibility also means that any defence of personal privacy must also take into account moderate group privacy, for affecting the latter does mean affecting the personal privacy of its members. I shall return to this point in the conclusion, where I will argue that even a moderate approach to group privacy requires taking the latter seriously in terms of legislation, in order to protect the privacy of the individual persons involved. If privacy applies to groups also in the strong sense seen above (recall also the analogy with the pile of books, which is heavy despite the fact that each book is light), then there is interest in exploring not only its consequences but also its nature, and this leads me to a final set of considerations.

---

[5] California S.B. 568 amends Division 8 of the California Business and Professions Code to add Chapter 22.1, see http://goo.gl/ODqtcO

5.5 What Kind of Privacy Can Group Privacy Be?

It is hard to elucidate the nature of group privacy—now understood in the strong sense clarified above—without a clear idea of what theory of privacy one is endorsing in the first place. Two theories are particularly popular in the current literature: the reductionist interpretation and the ownership-based interpretation. Neither is entirely satisfactory,[6] so I shall suggest a third one, based on the identity-constitutive nature of privacy, and argue that it is more suitable to understand strong group privacy.

The reductionist interpretation argues that the value of privacy rests on a variety of undesirable consequences that may be caused by its breach, either personally, such as distress, or socially, such as unfairness. Privacy is a utility, also in the sense of providing an essential condition of possibility of good human interactions, by preserving human dignity or by guaranteeing political checks and balances, for example.

The ownership-based interpretation argues that informational privacy needs to be respected because of each person's rights to bodily security and property, where 'property of *x*' is classically understood as the right to exclusive use of *x*. A person is said to own his or her information (information about him- or herself) and therefore to be entitled to control its whole life cycle, from generation to erasure through usage.

The two interpretations are not incompatible, but they stress different aspects of the value of privacy. The reductionist interpretation is more oriented towards a consequentialist assessment of privacy, in terms of cost–benefit analyses of its protection or violation. The ownership-based interpretation is more oriented towards a 'natural rights' understanding of the value of privacy itself, in terms of private or intellectual property. Unsurprisingly, because they both belong to a pre-digital culture, they both compare privacy breach to physical trespass or unauthorised invasion of, or intrusion in, a metaphorical space or sphere of personal information, the accessibility and usage of which ought to be fully controlled by its owner and hence kept private. As I have argued elsewhere (Floridi 2013, 2014), neither interpretation is entirely satisfactory in many respects.

The reductionist interpretation defends the need for respect for privacy in view of the potential misuse of the information acquired. So it is certainly reasonable, especially from a consequentialist perspective, to extend it to groups. However, it seems to support at most a moderate interpretation of group privacy; and recall that this is interesting only in terms of consequences. If all we are arguing is that groups may enjoy some privacy only because their members do, any reference to group privacy is a mere shortcut. Furthermore, the reductionist interpretation may be inconsistent with pursuing and furthering social interests and welfare. Although it is obvious that some public personal information may need to be protected—espe-

[6] See (Floridi 2013, 2014), for a detailed criticism, which is only summarized here insofar as it is relevant to the thesis defended in this chapter.

cially against profiling or unrestrained electronic surveillance—it remains unclear, on a purely reductionist basis, whether a society devoid of any privacy may not be a better society after all, with a higher, common welfare. Indeed, it has been convincingly argued[7] that the defence of privacy in the home—that is, within that special group represented by a family—may actually be used as a subterfuge to hide the dark side of privacy: domestic abuse, neglect, or mistreatment. Precisely because of reductionist-only considerations, even in democratic societies we tend to acknowledge that the right to privacy can be overridden when other concerns and priorities, including public safety or national security, become more pressing. All this by putting some significant interpretative pressure on the "arbitrary" clause that qualifies article 12 of The Universal Declaration of Human Rights which states that

> No one shall be subjected to *arbitrary* [emphasis added] interference with his privacy, family, home or correspondence, nor to attacks upon his honour and reputation. Everyone has the right to the protection of the law against such interference or attacks.

The ownership-based interpretation also falls short of being entirely satisfactory, for at least three reasons. First, informational contamination may undermine passive informational privacy. This is the unwilling acquisition of information or data, including mere noise, imposed on someone by some external source. Brainwashing may not occur often, but junk mail, or the case of a person chatting loudly on a phone nearby, are unfortunately common experiences of passive privacy breach, yet no informational ownership seems to be violated. Second, there is a problem of privacy in public contexts. Privacy—and especially group privacy, if there is such thing—is often exercised publicly, that is, in spaces that are socially, physically, and informationally shared: anyone can see what an individual person or group is doing downtown. How could a CCTV system be a breach of an individual's privacy if the individual in question is accessing a space that is public in all possible senses anyway? The ownership-based interpretation cannot provide a satisfactory answer. And finally, there is a metaphorical and imprecise use of the concept of 'information ownership', which cannot quite explain the lossless acquisition or usage of information. Information is not like a pizza: contrary to other things that one owns, one's personal information is not lost when acquired by someone else. Analyses of privacy based on 'ownership' of an 'informational space' are metaphorical twice over. All these difficulties make it less usable as a theory of group privacy. We need a better alternative, so here is a proposal.

Both the reductionist and the ownership-based interpretation fail to acknowledge the significant changes brought about by digital ICTs. They belong to an industrial culture of material goods, mechanical interactions, and of manufacturing/trading relations, so they rely on conceptual frameworks that are overstretched when trying to cope with the new challenges offered by an informational culture of services, networks, and usability. Interestingly, in their classic article The Right to Privacy,

[7] See (Fineman and Mykitiuk 1994), and especially the chapter by Elizabeth M. Schneider 'The Violence of Privacy' a reprint of her article published in 1990.

published in the Harvard Law Review in 1890, Samuel D. Warren and Louis Brandeis had already realised this limit with impressive insight:

> where the value of the production [of some information] is found not in the right to take the profits arising from publication, but in the peace of mind or the relief afforded by the ability to prevent any publication at all, *it is difficult to regard the right as one of property, in the common acceptation of the term* [emphasis added]. (Warren and Brandeis 1890), p. 25.

More than a century later, privacy requires a radical re-interpretation. Such a re-interpretation is achieved by considering each *individual person or group* as constituted by his, her or its information, and hence by understanding a breach of an individual's informational privacy as a form of aggression towards that individual's identity. This interpretation of privacy as having an identity-constituting value is consistent with the fact that ICTs can both erode and reinforce informational privacy, and hence that a positive effort needs to be made in order to support not only Privacy Enhancing Technologies but also constructive applications, which may allow users to design, shape, and maintain their identities as informational agents. The value of privacy is both to be defended and enhanced.

The information flow needs some friction in order to keep firm the distinction between the macro multi-agent system (the society) and the identity of the micro multi-agent systems (the individual persons and groups) within it. Any society (even a utopian one) in which no informational privacy is possible is one in which no identity-constituting process can take place, no personal or group identity can be developed and maintained, and hence no welfare can be achieved, social welfare being only the sum of the individuals involved. The total 'transparency' of the infosphere that may be advocated by some reductionists achieves the protection of society only by erasing all identity and individuality, a 'final solution' for sure, but hardly one that the individuals themselves, constituting the society so protected, would be happy to embrace. The advantage of the identity-constituting interpretation of privacy over the reductionist one is that consequentialist concerns may override respect for privacy, whereas the identity-constituting interpretation, by equating its protection to the protection of individual identity, considers it a fundamental right. By default, the presumption should always be in favour of its respect, even when we admit that privacy may be negotiable to some degree in special circumstances.

Looking at the nature of an individual person or group as being constituted by that individual's information enables one to understand the right to privacy as a right to immunity from unknown, undesired, or unintentional changes in one's own identity as an informational entity, both actively and passively. Actively, because collecting, storing, reproducing, manipulating etc. Alice's or her family's information amounts now to stages in cloning and fixing (profiling) their identities. Passively, because breaching Alice's or her family's privacy may now consist in forcing the individual or her group to acquire unwanted information, thus altering their nature as informational entities without consent. The first difficulty facing the ownership-based interpretation is thus avoided.

The identity-constituting interpretation suggests that a group's informational sphere and the identity of a group are co-referential, or two sides of the same coin.

The right to privacy, both in the active and in the passive sense just seen, shields the group's identity. This is why privacy is extremely valuable and ought to be respected. The second problem affecting the ownership-based interpretation is therefore also solved because violations of informational privacy are now more fruitfully compared to kidnapping rather than trespassing. The advantage, in this change of perspective, is that it becomes possible to dispose of the false dichotomy qualifying privacy in public or in private contexts. Some information constitutes a group context-independently, and therefore a group is perfectly justified in wishing to preserve its integrity and uniqueness even in entirely public places. Trespassing makes no sense in a public space, but kidnapping (even of a whole group) is a crime independently of where it is committed.

As for the third problem, one may still argue that an individual group 'owns' its information, yet no longer in the metaphorical sense seen above, but in the precise sense in which a group is its information. 'Its' in 'its information' is not the same 'its' as in 'its land' but rather the same 'its' as in 'its memories', 'its culture', 'its choices', 'its rites and customs', and so forth. It expresses a sense of constitutive belonging, not of external ownership, a sense in which its information is part *of* it but is not a (legal) possession *by* it. Once again, it is worth quoting Warren and Brandeis, this time at length, even if they had in mind the individual person, rather than an individual group:

> [...] the protection afforded to thoughts, sentiments, and emotions [...] is merely an instance of the enforcement of the more general right of the individual to be let alone. It is like the right not to be assaulted or beaten, the right not to be imprisoned, the right not to be maliciously persecuted, the right not to be defamed [or, the right not to be kidnapped, my addition]. In each of these rights [...] there inheres the quality of being owned or possessed and [...] *there may be some propriety in speaking of those rights as property. But, obviously, they bear little resemblance to what is ordinarily comprehended under that term.* The principle [...] is in reality not the principle of private propriety but that of *inviolate personality* [emphasis added]. [...] the right to privacy, as part of the more general right to the immunity of the person, [is] *the right to one's personality* [emphasis added].

This identity-constituting conception of privacy and its value has started being appreciated by more mature, information societies, where the identity-constituting interpretation reshapes some of the assumptions behind a still 'industrial', 'modern', or 'Newtonian' conception of privacy. The following considerations illustrate such a transition.

If some information is finally acknowledged to be a constitutive part of personal and group identity, then one day it may become strictly illegal to trade in some kinds of information, exactly as it is illegal to trade in human organs (including one's own) or slaves. At the same time, we might relax our attitude towards some kinds of 'dead individual information' that, like 'dead pieces of oneself', are not really, or no longer, constitutive of a person or a group. Legally, Alice may not sell her kidney, but she may sell her hair or be rewarded for giving blood. Likewise, her family may not sell its members, even if they all, unanimously, accept such a practice, but it may sell the properties of one of its deceased members as a group.

We are constantly leaving behind a trail of data, pretty much in the same sense in which we are shedding a huge trail of dead cells. The fact that nowadays ICTs allow our data trails to be recorded, monitored, processed and used for social, political or commercial purposes is a strong reminder of our informational nature as individual persons and groups. It might be seen as a new level of environmentalism, as an increase in what is recycled and a decrease in what is wasted (not unlike what bacteria do with DNA available in the environment). At the moment, all this is just speculation and in the future it will probably be a matter of fine adjustments of ethical sensibilities, but the third Geneva Convention (1949) already provides a clear test of what might be considered 'dead personal information'. A prisoner of war need only give his or her name, rank, date of birth, and serial number and no form of coercion may be inflicted on him or her to secure any further information, of any kind. Even if we were all treated fairly as 'prisoners of the information society', our privacy would be well protected and yet there would still be some personal data that would be perfectly fine to share with any other agent, even hostile ones. It is not a binary question of all or nothing, but an analogue one of fine balance and degree.

A further issue that might be illuminated by looking at privacy from an identity-constituting perspective are those of confidentiality and intimacy, two intrinsically group-based phenomena. The sharing of private information with someone, implicitly (especially by doing things together), or explicitly, through communication, is based on a relation of profound trust that binds the people involved intimately. This coupling is achieved by allowing persons to be partly constituted as selves by the same information. The union of the persons in question forms a single unity, a supra-agent, or a new multi-agent individual, the group. Precisely because entering into a new supra-agent is a delicate and risky operation, care should be exercised before 'melding' oneself with other individuals by sharing personal information or its source, such as common experiences. This is the way I interpret the concluding sentence of *The Catcher in the Rye*, the famous novel by J. D. Salinger:

> Don't tell anybody anything. If you do, you start missing everybody. (Salinger 1951)

Confidentiality and intimacy create a bond that is hard and slow to forge properly, yet resilient to many external forces when finally in place, as the group (the supra-agent) is stronger than the constitutive agents themselves. Relatives, friends, classmates, fellows, colleagues, comrades, companions, partners, teammates, spouses and so forth may all have experienced the nature of such a bond, the stronger taste of a 'we'. But it is also a bond that is brittle and difficult to restore when it comes to internal betrayal, since the disclosure, deliberate or unintentional, of some personal information in violation of confidence can entirely and irrecoverably destroy the intimacy and privacy of the new, supra-agent born out of the joining agents, by painfully introducing discord. The 'we' is strongly armoured against 'the other', but extremely fragile against internal betrayal by 'one of us'.

A final issue can be touched upon rather briefly: the identity-constituting interpretation stresses that privacy is also a matter of construction of an individual's own identity. The right to be left alone is also the right to be allowed to experiment with one's own life, to start again, without having records that mummify one's personal identity forever, taking away from the individual person or group the power to form and mould who or what the individual is and can be. Every day, an individual person or group may wish to build a different, possibly better, 'I' or 'we'. We never stop becoming ourselves, so protecting persons and group privacy also means allowing that person and group the freedom to construct and change herself or itself profoundly. The right to privacy is also the right to a live, renewable identity that one can shape freely. This is why it matters.

5.6 Conclusion

The idea that groups may have (at least something akin to) a right to privacy is not new (see for example (Bloustein 1978, 2003)) and it is open to debate (Bisaz 2012). But it has not received the attention it deserves, although the issue is becoming increasingly important. And this because, by far, ICTs treat most people not as individuals but as members of specific groups (or classes, collections, crowds, populations and their segments etc.), where the groups are the really interesting focus, as carriers of rights, values, and potential risks. Think of the owners of such and such kind of car, shoppers of such and such kinds of goods, people who like this type of music, or people who go to that sort of restaurant, cat owners, dog owners, people who live in a specific postal code, carriers of a specific gene, people affected by a particular disease, team fans … Especially big data is more likely to treat types (of customers, users, citizens, demographic population, etc.) rather than tokens (you, Alice, me…), and hence groups rather than individuals. But re-identifiable groups are *ipso facto* targetable groups. And membership in a sufficient number of groups can easily lead to the re-identification of individuals. Indeed, in terms of logic, two sets (even if they are infinite) are already sufficient to identify a singleton (a set with exactly one element). As an elementary example, suppose A is the infinite set of all integers including and larger than 1, and B is the infinite set of all integers including and smaller than 1, their intersection contains exactly one element, namely 1 ($A\ B = 1$). It is therefore a very dangerous fallacy to think that, if we protect personal data that identify people individually, the protection of groups of people will take care of itself. I have argued above that we should consider group privacy as something that is sometimes reducible to the individual privacy of its members, and sometimes as something that belongs to the group as a group. I have defended the plausibility of both moderate and strong group privacy. But I have also stressed that defending moderate group privacy is already crucial, in terms of the significant nature of its consequences. This is not the current view. In particular, a 'nominalist' approach (or informational ontology (Floridi 2003)) to group privacy—take care of each member separately and the group will automatically be fine too—is currently at the roots of European legislation. This defines a "Data Subject" as:

> An identified or identifiable person to whom specific personal data relates. It is someone who can be identified, directly or indirectly, in particular by reference to an identification number or to one or more specific factors (physical, physiological, mental, economic, cultural, social). (European commission).

As a consequence, both the 1995 Directive and the new Regulation under discussion focus on individual persons. The philosophy informing the approach may be grasped by looking at the following recitals (emphases added):

> Whereas the principles of protection must apply to any information concerning an identified or identifiable *person*; whereas, to determine whether a *person* is identifiable, account should be taken of all the means likely reasonably to be used either by the controller or by any other person to identify the said *person*; whereas the principles of protection shall not apply to data rendered anonymous in such a way that the *data subject* is no longer identifiable […]. (Directive 95/46/Ec)

and, even more restrictively (notice the "natural"):

> The principles of protection should apply to any information concerning an identified or identifiable *natural* person. To determine whether a *natural* person is identifiable, account should be taken of all the means likely reasonably to be used either by the controller or by any other person to identify the individual. The principles of data protection should not apply to data rendered anonymous in such a way that the data subject is no longer identifiable. (Com(2012) 10 final 2012/0010 (Cod)).

Yet even from a nominalist perspective, we should acknowledge that both friendly and hostile users of big data may not care about Alice at all, but only about the fact whether Alice, whoever she is, belongs to the group that regularly goes to the local church, or mosque, or synagogue, uses Grindr, or has gone to a hospital licensed to carry out abortions, or indeed shares a feature of your choice. In military terminology, Alice is hardly ever a High Value Target, like a special and unique building. She is usually part of a High Pay-off Target, like a tank in a column of tanks. It is the column that matters.

As I have argued elsewhere (Floridi 2013) our current ethical approach is too anthropocentric (only natural persons count) and nominalist (only the single individual person counts). We should take other kinds of individuals, including groups, into account. We need to be more inclusive because we are underestimating the risks involved in opening anonymised personal data to public use, in cases in which *groups* of people may still be easily identified and targeted. Such inclusiveness should not be too hard to achieve. After all, we already accept as ordinary the fact that groups as *agents* may infringe on someone's privacy. In the United States, we are used to considering as normal collective lawsuits (class actions) in which a group may sue a person or another group. And in Europe, consumer organisations regularly bring claims on behalf of the groups they represent. Clearly, there are cases in which the protection of a right requires a balance between the agents, issuing the action, and the patients, receiving the action.

There are very few Moby-Dicks. Most of us are sardines. The individual sardine may believe that the encircling net is trying to catch it. It is not. It is trying to catch the whole shoal. It is therefore the shoal that needs to be protected, if the sardine is to be saved. An ethics addressing each of us as if we were all special Moby-Dicks may be flattering and perhaps, in other respects, not entirely mistaken, but needs to be urgently upgraded. Sometimes the only way to protect a person is to protect the group to which that person belongs. Preferably before any disaster happens. This moderate sense of group privacy is the least we should begin to consider, as a first step towards a full recognition of strong group privacy.

Acknowledgements I am very grateful to Massimo Durante for his most valuable comments on a previous draft of this article; to Ugo Pagallo and the participants in the panel "Open Data and Data Protection: Problems and Perspectives"—organised at the conference *Computers, Privacy and Data Protection 2014 (CPDP 2014) Reforming Data Protection: The Global Perspective*—for their feedback; to the participants in the workshop on group privacy held in Amsterdam on the 8th of September 2014 for valuable discussions; to David Sutcliffe for his skilful copyediting of the final version and many insightful comments that improved it significantly; and to Linnet Taylor for her feedback on a penultimate draft of this chapter, and some enlightening conversations on the topic of group privacy. Her chapter in this volume makes a strong and convincing case for the protection of group privacy in contexts of geolocated data, especially in LMICs (see also (Taylor 2016)).

Bibliography

Beebee, H., and N. Sabbarton-Leary. 2010. *The semantics and metaphysics of natural kinds, Routledge studies in metaphysics*. New York/London: Routledge.

Bisaz, C. 2012. *The concept of group rights in international law: Groups as contested right-holders, subjects and legal persons*. Leiden: Brill, Nijhoff.

Bloustein, E.D.J. 1978. *Individual and group privacy*. New Brunswick: Transaction Publishers.

Bloustein, E.J. 2003. *Individual & group privacy*, 2nd ed. New Brunswick: Transaction Publishers.

Campbell, J.K., M. O'Rourke, and M.H. Slater. 2011. *Carving nature at its joints: Natural kinds in metaphysics and science, topics in contemporary philosophy*. Cambridge/London: MIT Press.

COM. 2012. 10 Final 2012/0010 (COD). Proposal for a Directive of the European Parliament and of the Council on the Protection of Individuals with Regard to the Processing of Personal Data by Competent Authorities for the Purposes of Prevention, Investigation, Detection or Prosecution of Criminal Offences or the Execution of Criminal Penalties, and the Free Movement of Such Data/* Com/2012/010 Final – 2012/0010 (Cod) */

Directive 95/46/EC. Directive 95/46/Ec of the European Parliament and of the Council of 24 October 1995 on the Protection of Individuals with Regard to the Processing of Personal Data and on the Free Movement of Such Data.

European Commission. Justice, data protection, glossary, data subject: Http://Ec.Europa.Eu/Justice/Data-Protection/Glossary/Index_En.Htm.

Fineman, M.A., and R. Mykitiuk (eds.). 1994. *The public nature of private violence: The discovery of domestic abuse*. London/New York: Routledge.

Floridi, L. 2003. Informational realism. In *Selected papers from conference on computers and philosophy-Volume 37*,eds. J. John Weckert and Y. Al-Saggaf, 7–12. Australian Computer Society.

Floridi, L. 2013. *The ethics of information*. Oxford: Oxford University Press.

Floridi, L. 2014. *The fourth revolution – How the infosphere is reshaping human reality*. Oxford: Oxford University Press.

Groves, P., B. Kayyali, D. Knott, S., and van Kuiken. 2013. The 'big data' revolution in healthcare. *McKinsey Quarterly*.

Howe, D., M. Costanzo, P. Fey, T. Gojobori, L. Hannick, W. Hide, D.P. Hill, R. Kania, M. Schaeffer, and S. St Pierre. 2008. Big data: The future of biocuration. *Nature* 455(7209): 47–50.

Khalidi, M.A. 2013. *Natural categories and human kinds: Classification in the natural and social sciences*. Cambridge: Cambridge University Press.

LaPorte, J. 2004. *Natural kinds and conceptual change*. Cambridge: Cambridge University Press.
Mittelstadt, B., and L. Floridi. 2016. The ethics of big data: Current and foreseeable issues in biomedical contexts. *Science and Engineering Ethics* 22(2): 303–341.
Mittelstadt, B., Floridi, L. eds. forthcoming-b. *The ethics of biomedical big data, law, governance and technology*. New York: Springer.
Oderberg, D.S. 2013. *Classifying reality*. Chichester: Wiley-Blackwell.
Panchen, A.L. 1992. *Classification, evolution, and the nature of biology*. Cambridge: Cambridge University Press.
Richards, R.A. 2010. *The species problem: A philosophical analysis*. Cambridge: Cambridge University Press.
Salinger, J.D. 1951. *The catcher in the rye*. London: H. Hamilton.
Taylor, L. 2016. No place to hide? The ethics and analytics of tracking mobility using mobile phone data. *Environment and Planning D: Society and Space* 34(2): 319–336.
Warren, S., and L. D. Brandeis. 1890. The right to privacy. *Harvard Law Review* 193(4).

Chapter 6
Social Machines as an Approach to Group Privacy

Kieron O'Hara and Dave Robertson

Abstract This chapter introduces the notion of social machines as a way of conceptualising and formalising the interactions between people and private networked technology for problem-solving. It is argued that formalisation of such 'social computing' will generate requirements for information flow within social machines and across their boundaries with the outside world. These requirements provide the basis for a notion of group privacy that is neither derivative from the idea of individual privacy preferences, nor founded in political or moral argument, but instead related to the integrity of the social machine and its capabilities for bottom-up problem-solving. This notion of group privacy depends on a particular technological setup, and is not intended to be a general definition, but it has purchase in the context of pervasive technology and big data which has made the question of group privacy pressing and timely.

Keywords Social machines • Social computing • Group privacy • Identity • Collective action • Social networking

6.1 Introduction: Technology and the Ideology of Privacy

Group privacy is an interesting topic made more salient in recent years by the growth of big data, enabling people to be targeted and understood via their personal attributes (on the basis of correlations between the people who possess those attributes and exogenous phenomena), or alternatively via the properties of their networks (for instance, one's social network is a strong predictor of whether one is likely to default on a loan – cf. e.g. Seiler et al. 2011). In each case, the dilemmas of

---

K. O'Hara (✉) • D. Robertson
Southampton University, Southampton, UK
e-mail: kmo@ecs.soton.ac.uk

L. Taylor et al. (eds.), *Group Privacy*, Philosophical Studies Series 126,
DOI 10.1007/978-3-319-46608-8_6

privacy are thrown into sharp relief – visibility to one's network brings benefits, but compromises privacy.

Furthermore, there is a distinct potential for injustice, as one may find oneself discriminated against on the basis of behaviour of *other* people in one's groups (Hildebrandt 2012). The injustice and the privacy are different phenomena which are likely to require separate consideration, but there is a *prima facie* case for arguing that we could nip the injustice in the bud if groups as well as individuals had privacy rights. In the age of big data, data crunchers are not interested in the individual data points, so much as the mass (Floridi 2014) – yet the crunching of data about the mass can and does have real-world implications for individuals.

This is one of the many ways in which data protection is an imperfect protection for privacy (O'Hara 2011, 7–11). Data protection requires an individual to be identifiable before data is classified as personal data, so that the subject's consent is required for processing (of course there are many exceptions to this built into data protection legislation). Yet the notion of group privacy is consistent with something that we intuitively understand in the age of spam, junk mail and racial profiling – one need not be identified to have one's privacy invaded. The mere existence of a non-identifying profile of oneself, combined with a point of access such as an address, may not count as personal data, but is still an annoying invasion.

Couching the problem in this way still leaves the privacy of the group derivative from the privacy of the individual – the individual's remedy for the invasion of his personal privacy is to insist on the privacy of a wider group of which he is an anonymous member. This seems to chime in with liberal ideas about privacy; a major contribution of privacy to social value, according to one influential analysis, is to support individual autonomy (Rössler 2005). Meanwhile, intrusion from the group itself has, since Mill, been seen as a serious threat to the individual (Mill 1859). Some theorists, for example feminists, have argued that the privacy of small units (from the family upwards) is a means of concealing abuse rather than of legitimately supporting the individual (MacKinnon 1989; Allen 2003).

In the liberal tradition that often conceptualises the group as a threat to the individual, group privacy does not look like a serious runner, *unless* the group can be reconceptualised as an important support for the individual's autonomy – and so group privacy seems to derive its value from the needs of the individual, not the group itself.

On the other hand, a conservative viewpoint is more ambivalent about the power and potential of an individual – for instance, Burke lauds the little platoons, and considers the individual as intrinsically unable to make consistent or wise moral judgments. Schoeman argues, *contra* Mill, that social control, far from being morally destructive, is an important factor in a valuable liberty. Our competence as rational agents depends on constructive adaptations of social control mechanisms in real-world contexts. Unpicking informal social control mechanisms in the name of autonomy, in Schoeman's view, actually deprives the individual of important social abilities, and "helps maintain both the integrity of intimate spheres as against more public spheres and the integrity of various public spheres in relation to one another" (Schoeman 1992, 157).

Adam Smith's view, with regard to the moral education of people in the newly emerging metropolises of the eighteenth century, is an interesting example of this kind of thought.

> A man of low condition, on the contrary, is far from being a distinguished member of any great society. While he remains in a country village his conduct may be attended to, and he may be obliged to attend to it himself. In this situation, and in this situation only, he may have what is called a character to lose. But as soon as he comes into the great city, he is sunk in obscurity and darkness. His conduct is observed and attended to by nobody, and he is therefore very likely to neglect it himself, and to abandon himself to every sort of profligacy and vice. (Smith 1994, vol.2, V.i.g.12, 795, footnote omitted)

The way to address this, thought Smith, was not more policing or the reduction of the private sphere of the 'man of low condition', but rather greater power for groups, specifically those that have an interest in the individual's moral conduct.

> He never emerges so effectually from this obscurity, his conduct never excited so much the attention of any respectable society, as his becoming the member of a small religious sect. He from that moment acquires a degree of consideration which he never had before. All his brother sectaries are, for the credit of the sect, interested to observe his conduct, and if he gives occasion to any scandal, if he deviates very much from those austere morals which they almost always require of one another, to punish him by what is always a very severe punishment, even where no civil effects attend it, expulsion or excommunication from the sect. In little religious sects, accordingly, the morals of the common people have been almost always remarkably regular and orderly; generally much more so than in the established church. (Smith 1994, vol.2, V.i.g.12, 795–6).

In this chapter, we shall attempt to pick a way between these two ideologically-charged interpretations of group privacy, to suggest one potential characterisation applicable to the data-heavy online world which has made the question of group privacy appear so pressing. This characterisation is intended to be entirely technical in nature, and independent of the question of whether the privacy (or indeed the integrity) of the group is a good thing. Such moral and legal questions cannot be ducked, but it may make them more easily addressable if questions of the existence, nature and effect of privacy can be resolved separately.

In the next section, we will argue that some such characterisation of the problem as our own is necessary, in the big data era, to make sense of group privacy. Then we will introduce the idea of social machines, and in the following section consider how we might use them to understand privacy concerns. The next section will sketch an abstract characterisation of social machines and social computing, to give a sense of how privacy concerns may be discovered. This abstract specification is given a little more flesh with some examples in the next section, before we discuss privacy aspects in a little more detail.

6.2 Complexity, Identity and Big Data

Would group privacy create greater complexity in policing and vigilance, and would it be a right going beyond existing expectations, preferences and the needs of democratic societies? Individual privacy introduces a number of private spaces proportional (of course) to the number of citizens, whereas group privacy will be a correspondingly complex concept to enforce.

If we think about the number of groups that people are likely to claim they are members of, and whose corporate privacy they wish to defend, the extra complexity grows in a linear fashion as population grows. On average, people might admit to membership of m groups (m maybe between 10 and 100), while average membership of a group would be n people. Hence, for a population of x, the number of groups to be protected would be proportional to mx/n.

However, big data will change this. Data mining finds significance in correlations between people with no obvious connection, or put another way within groups that have no external significance. One might easily not know, or care, that one was a member of such a group (such as, for instance, 26–35 year old males earning between £40 k–£50 k p.a. in households without children who have downloaded more than 5 unsolicited recommendations from Spotify in the last 6 months). Even if we adopt an extensional characterisation for a group (which may not be the best way of characterising groups), for a population of x the number of such potential groups is 2^x-1, but as big data crunchers do not consider the coherence or independent interest of such groups it would be hard to single out which groups are worth protecting. This could create an extremely complex and difficult legal scene, with hard decisions to make about liability and the balance between social good and protection of rights.

So it is impractical to consider theoretically possible groups, whose number will grow exponentially with the population. The monitoring and policing of group privacy can more easily be kept tractable if we take into account those groups that individuals expressly understand themselves to be members of. In that sense, group privacy will remain derivative from individual privacy, but crucially in this case the value of the group's privacy can be decoupled from the individual's privacy if the group can be instrumented to detect the effects on the group itself of different privacy strategies.

Participation in groups helps cultivate certain values and virtues in the members. Which ones are cultivated depends somewhat on the nature of the group in question. Membership tends to create individuals who are predisposed to internalise, uphold and perpetuate the values and virtues of that environment. This is what Nancy Rosenblum (2000) has termed the 'logic of congruence'. Adam Smith believed that this was inherently valuable as a means of socialising individuals, and privacy may be the sort of value that could be cultivated in this way.

Given that view, it may make sense to look at group privacy as a means of empowering the group to achieve its aims, and to see its protection as a means of institutionalising that empowerment. Of course this does not resolve the ethical question of when that is a good thing and when bad, but at least it gives us a rationale to do it. In the big data context, we should consider whether current conceptualisations of technology give us a handle on group empowerment. With that in mind, we now introduce the topic of *social machines*.

6.3 Social Machines

The world of big data has not, of course, been unaccompanied by other developments. In particular, as the amount of data that it is feasible to process has grown, so has the number of people that it is feasible to connect within a network. Figure 6.1, following David De Roure, gives a sense of different interaction modes of computing. Where there are more machines, to produce the big data paradigm in the upper left, or more people, as in the social networking paradigm in the lower right, distribution of computational resources is inevitable, and hence Web or Web-like technologies are necessary to handle the interaction at scale. The technological affordances have, over time, moved upwards and towards the right, ultimately to reach the fourth quadrant.

As the number of people and machines linked together grows, and as the intelligence of the machines increases, we can treat goal-driven networks as individual systems, or *social machines* (Berners-Lee 1999; Hendler and Berners-Lee 2010; Shadbolt et al. 2013; O'Hara et al. 2013, 2014; De Roure 2014), a nascent focus of computing research (Bernstein et al. 2012). 'Programming the global computer' or 'global ubiquitous computing' has been recognised as a grand challenge for computing (Kwiatkowska et al. 2004), and now the technologies of software agents (Jennings et al. 2014) and peer-to-peer technologies flexibly link people and computers, as explored in projects such as SOCIAM (http://sociam.org/), OpenKnowledge (http://www.openk.org/) and the Social Computer community (http://www.social-computer.eu/). As we unravel the mysteries of scale and control, we will need not just to understand the emergent phenomena, but to develop means, methods and tools for controlling them, at least partially (O'Hara et al. 2013). The problem is sharpened by the desideratum that 'programming the social computer' must be achievable from *within* the social computer – research should democratise control by supporting people in the development of social machines to achieve their own smaller-scale, local, idiosyncratic purposes.

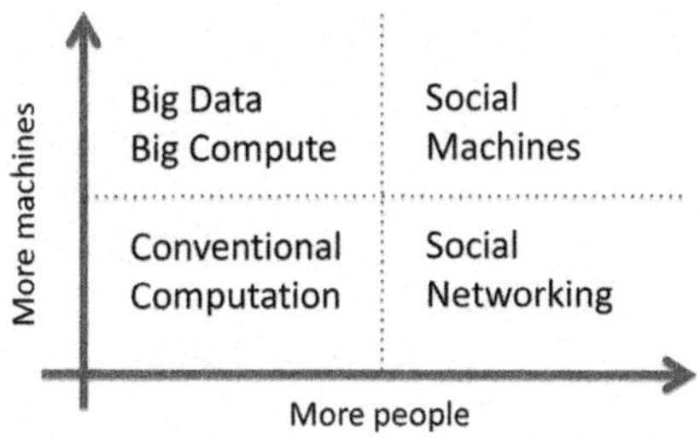

Fig. 6.1 A matrix showing the affordances of scale (Adapted from De Roure 2014)

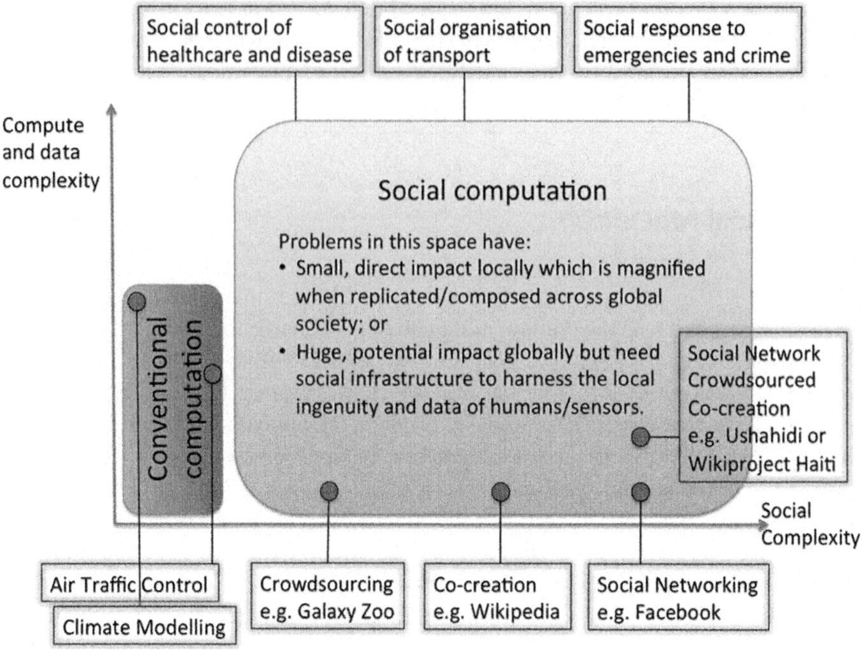

Fig. 6.2 The space of social machines (O'Hara et al. 2013)

Figure 6.2 shows the potential space in more detail. Conventional computation, even in highly complex domains such as air traffic control and climate modelling, appears on the left hand side, where social complexity is low even if computational complexity is high. Crowdsourcing systems, such as the citizen science initiative Galaxy Zoo (Lintott et al. 2008) have a relatively low level of social complexity as well. Conversely, even systems with high social complexity still currently involve relatively low computational complexity. More complex social arrangements are required for the co-creation of content, e.g. Wikipedia, and social networking. However, when these interactions combine, where a social network acts as a platform for crowdsourced co-creation of content, as recently happened with the Ushahidi map of election violence in Kenya in 2007 (Okolloh 2009), or the reuse of Ushahidi to create a post-earthquake map of Port-au-Prince in Haiti in 2010 (Morrow et al. 2011), we start to see more complex interactions emerging between people in the 'social computer' and their environment. As we explore this *terra incognita* of social computation, in order to address issues with collective action problems, such as public health, transport or crime, we would expect to find solutions with small impacts locally being magnified at scale, as long as the requisite infrastructure (including Web technology) is in place.

The idea of a social machine has been implicit throughout the history of the Web. As Berners-Lee put it in 1999:

> Real life is and must be full of all kinds of social constraint – the very processes from which society arises. Computers can help if we use them to create abstract *social machines* on the Web: processes in which people do the creative work and the machine does the administration. (Berners-Lee 1999, 172, Berners-Lee's emphasis)

Many social machines are built on social networking sites such as Facebook, in which human interactions from organising a birthday party to interacting with a Member of Parliament are underpinned by the engineered environment. Another type of example is a multiplayer online game, where a persistent environment facilitates interactions concerning virtual resources between real people. A different type of game is online poker, where the resources being played for are real-world, where the players may be human or bots, and where the environment in which the game takes place is engineered around a relatively simple computational model. In such systems, (some of) the social constraints that Berners-Lee talks about, currently norm-driven, are administered by the architecture of the programmed environment.

A generalised definition of a social computation is provided by (Robertson and Giunchiglia 2013):

> A computation for which an executable specification exists but the successful implementation of this specification depends upon computer mediated social interaction between the human actors in its implementation.

In such an environment, self-organisation (partial or full) becomes viable and scalable, while physical objects, agents, contracts, agreements, incentives and other objects can be referred to using Uniform Resource Identifiers (such as Web addresses), thereby allowing consistent context-independent reference throughout the executable specification. 'Programming' the social computer (as opposed to simply supporting and directing interactions in an engineered environment) and integrating larger numbers of people and machines will become increasingly feasible.

As a small example of a social machine, consider reCAPTCHA (Von Ahn et al. 2008). A CAPTCHA (Completely Automated Public Turing test to tell Computers and Humans Apart), invented by Louis Von Ahn, is the distorted sequence of letters that someone has to type in a box to identify him- or herself as a human (e.g. to buy a ticket online, or to comment on a blog). This is a task that computers cannot do, and so the system stops bots buying thousands of tickets for a concert or sporting event for later resale, or for a spambot to leave spam messages as comments to blogs (Von Ahn et al. 2003).

Von Ahn extended the idea of the CAPTCHA to create the reCAPTCHA, which socialises the same principle to solve another problem. Google (which acquired reCAPTCHA in 2009) uses it to scan older books automatically. The original CAPTCHA device was being used over 200 m times a day, about half a million person-hours of effort. reCAPTCHA puts these person-hours to more productive use, presenting the user who wishes to identify him- or herself as a human with two words, not one. The first is a normal CAPTCHA, and the second is a word from an old book that Optical Character Recognition had failed to identify. If the person succeeds with the first CAPTCHA, then he or she is known to be a human. As humans are reliable at word recognition, the response to the second word as a plausible suggestion of what it is. Presenting the same word to multiple users allows a consensus to emerge. The goal of the social machine is to digitise books – people's needs to prove themselves human provides the mechanism.

However, reCAPTCHA is purely exploitative, as the goal of the machine is independent of the requirements of its human 'components'. As another example, Robertson and Giunchiglia 2013 use the DARPA balloon challenge of 2009, in which the aim was to find ten weather balloons placed randomly around the US (in nine different states from California to Delaware). The rules of the challenge were intended to support the growth of a network of people taking part in the search, enabling a crowdsourced solution. The means of doing this in the winning solution (from Sandy Pentland at the Massachusetts Institute of Technology) was to set out financial incentives according to a Query Incentive Network Model (Kleinberg and Raghavan 2005), in which people were incentivised both to look for the balloons and to add more people to the network. Pentland's team began with 4 people, and using social media had recruited over 5000 at the point of completion, which took under 10 h (Pickard et al. 2010).

reCAPTCHA and the DARPA challenge were top-down systems designed to solve a particular problem, but social machines can, and indeed should (O'Hara 2012), solve the problems of the people who constitute them. In such cases, the incentive of the participants is that the machine's smooth functioning is in their own interests. One could imagine, for instance, a set of computer-mediated interactions enabling a community to provide a social response to problems of crime (such as BlueServo, http://www.blueservo.net/, which crowdsources the policing of the Texas-Mexico border), or enabling those suffering from a particular health care problem to pool resources and to offer support and advice to fellow sufferers (such as curetogether.com, http://curetogether.com/). There is a growing number of health social machines, as surveyed in detail in (Van Kleek et al. 2013). It will be obvious from these examples, particularly BlueServo, that such efforts will not always be uncontroversial. Attempts to crowdsource the identities of the bombers of the Boston Marathon in 2013 bordered on farce, and, although the countercultural website 4chan was prominent in the homemade policing efforts with its so-called '4chan Think Tank', its lamentable efforts were soon parodied elsewhere on the same site (Walker 2013). Trust will be a major factor in the success of such machines (O'Hara 2012).

6.4 Social Machines as an Approach to Group Privacy

Suppose methods and tools were available to enable and empower communities to use data and networked communications to solve self-identified problems. In such a world – which is not yet in existence, although such tools are being actively researched – the social machine would be part of a community's repertoire of problem-solving resources. In that event, the social machine would have certain functional requirements which we could uncover by examining its 'program' – i.e. the computations that it would carry out to transform the state of the world – in abstract terms, independent of whether the computing was being done by machines, people or groups of people (Robertson et al. 2014). In the next section, we will sketch the sort of language that would serve this purpose.

Such an abstract specification would need to be 'filled in' by accounts of the key social and psychological factors for the machine to function – for instance, the decisions by actors to engage with the machine, the knowledge that actors bring to the computation, the restrictions on who can participate and so on. Modelling a socio-technical interaction in this way would include specifications of how information needs to flow around the machine, and to and from the machine as inputs and outputs.

For instance, one might discover that a particular computation $n(S_1, S_2)$, which transforms the world from state S_1 to state S_2 would only be possible if the actors involved, who carry out *n*, freely exchange personal data with each other, while ensuring that it does not spread beyond their circle. Or it might be that the actors require access to the personal data of all participants in the social machine (including of those not involved in that particular computational step). Or it might be that external services might be required which need access to personal data, or anonymised personal data, of participants in the social machine.

A statement in terms of the information-processing needs of the social machine would help make the demands made on information about the group as a whole explicit in terms of the goals which it wishes to achieve. In that way, we might find ways of explaining the functional value of privacy for a particular group, independently of moral generalisations about group privacy rights – prior to the moral question of whether a group should be empowered to achieve its particular goals.

Hence the social machine paradigm becomes a method of specifying what information requirements a particular mode of interaction creates. As noted above, privacy can be, and usually is, an issue for an individual. Thinking in terms of social machines, we can see how group privacy might be expressed, modelled or derived.

The definition of 'privacy' is of course highly contested (Solove 2006), and we don't want to enter that philosophical discussion in this paper. But if the reader will grant a basic – no doubt flawed and partial – definition of privacy as the ability of an individual to control access of others to her, including to information about her, then we can consider what a social machine specification will tell us about privacy within the context of that particular interaction.

The social machine will have requirements for information, including personal information. For instance, a map of political violence, such as the Kenyan Ushahidi project discussed above, will require photographs, and possibly eyewitness testimony. Crucially, the photographs would have to be stamped with location and time in their metadata, because these are the key dimensions. This information may well identify, or be disclosive about, the photographer, and we can assume, in this example about political violence, that a breach of privacy entails a level of risk for the individual. In this case, the photographer's privacy would be compromised beyond the social machine, as the photographs, dates and times would have to be displayed for the machine to function. It would be the photographer's decision, therefore, as to whether her lack of privacy in that case would create a risk of her being identified as a person in a specific place or time. In group terms, those taking part in the mapping exercise would agree to give up their privacy at least so far as revealing the photographs' metadata.

Let us go further, and imagine that the machine also needs a further input: an email address associated with the photograph. This could be for a number of reasons: it may be that the level of social trust in the society is so low that verification of identity is essential for the output to be usable; it may be that a number of false claims is too high and threatening to skew the output; it may be that one of the machine's functions is to generate follow-up information after a few hours or days to check on the progress of violence; it may be that the machine's accuracy depends on mechanisms to ensure that its account of the violence is taken from a representative group of actors. The email address is of course far more privacy-threatening than the time and location of the photograph – indeed, it could be directly identifying. Even if not, it could reveal, say, the workplace of the sender. However, there is no need as far as the machine is concerned to broadcast this extra information which would not enhance the actual map of the violence.

The social machine needs certain breaches of privacy to function. It does not need to know a participant's name or address – they can remain private to the participant. It needs a participant's email address, but does not need to publish this. Depending on the group's constitution, it may be that all group members need access to emails, or (more securely) only a small administrative cadre would have that access. It does need to publish the location and time of the photograph in the clear.

The participants form a group cooperating to produce the map of political violence, and that cooperation requires certain limits to individual privacy and makes certain demands about information flow. The cooperative project requires photographs, together with associated metadata of date, location and email address, otherwise it could not deliver its output. The group is prepared to release the location and time of the photographs, but it contains (or uses) other information – the email addresses and their association with specific photographs –which it is not prepared to release to the outside world. If there was a perception that information volunteered to the group might be made available to a wider public, then it could be anticipated that fewer people would be prepared to join the group, and that therefore the map would be less comprehensive.

Our claim, then, in this big data context, is that this is a notion of group privacy. The group's aim would be unachievable, or less achievable, if the information that it held was made available outside the group. The particular privacy regulation is a precondition for the group's success. It is not derivative from any individual privacy preferences – for example, individuals within the group will have varying preferences about this condition, and those who are particularly privacy conscious would no doubt work around the breach, by concocting misleading addresses, or sharing them with other group members. Others will be willing to be directly associated with their photographs. The group – and we imagine this at scale, so maybe millions of people – cooperates to achieve a goal which requires the publication of certain potentially identifying information, and the processing of other information which is not to leave the group.

Group privacy, then, can be considered in this context to be the requirement to keep information which is necessary/desirable for the operation of a social machine within the social machine. There would be various gradations of this; maybe such information should be available to all group members, or to some members with administrative privileges, or to those group members who need to cooperate on smaller subtasks, or should simply be available to an automatic processing module and not seen by any human being. Maybe it should be stored for a period of time, or deleted immediately, or maybe it doesn't matter. Maybe it should be anonymised. These are issues that would need to be addressed in the design or development of the system.

It is clear that such a characterisation is not one based on rights, claims or entitlement, and also does not include a notion of control. Rather, this is an understanding of privacy as a state of limited access – though in this case, not to a person, but to the group (cf. Schoeman 1984). The group's privacy requirements are derived from its goal, not from the privacy preferences (or rights) of its members. This is a means of deriving and expressing requirements – there is no normative claim being made that the machine should be allowed to operate. Rather, the social machine paradigm expresses a privacy requirement, which can then be measured against relevant social norms, regulations and other rights and protections to assess its desirability.

In this sense (and we accept there may be other perfectly reasonable senses of group privacy), group privacy is not derivative from the preferences of the social machine members. It might be that the functioning of the social machine will depend on its privacy arrangements being consistent with the preferences of the members; a social machine that was cavalier with its members' individual privacy would have no members. But the more important point here is that the privacy settings of the machine itself are determined by the requirements of the machine's function. If certain information within the machine is exported to the outside world, then the machine would cease to function. The machine's functioning is *not* derivative from the attitudes of individuals, but is rather *emergent* from their interactions.

As an example of how the sensitivity of privacy might apply only to the group and not to its members, consider a social machine, a variant of the violence mapping machine sketched out above, designed to map *and deter* non-violent anti-social behaviour (corruption perhaps – see Zinnbauer 2014 for examples). In that case, the social machine participants might individually be unconcerned if their identities as group members were known, if we assume they were in no physical danger. Indeed they might be admired by other members of the community for their work, so their preference might be to eschew their individual privacy and to reveal their identities. However, the social machine itself might function best as a deterrent if the identities of its participants, and the size of the group, were *not* known by government officials. In that case, there is a need for group privacy of group members' identities, based on the successful functioning of the social machine despite being against the wishes of the group members.

6.5 Specifying a Social Machine

What would such a high level, abstract specification of a social machine look like? In this section we will make some suggestions about such a program at a level abstracted away from immediate implementation concerns, therefore defining a class of systems. There could therefore be many different social machines all operating to the same specification with different combinations of people and computers. We will then consider an example application of social machine thinking in the area of healthcare.

6.5.1 Specification, Implementation and Non-computational Issues

We will present a specification language based on recursion for an abstract computer in as simple a way as possible (a more expressive lightweight formal language is presented in Robertson et al. 2014). Of course, as this computer includes human elements and social interactions, this specification will gloss over myriad complexities. However, as we shall argue, this is a feature of the language, not a bug. In this section, we will show how the logical characterisation of information flow will reveal the need for human/social characterisations of the social interactions involved.

The logical connectives used throughout are the standard ones for implication, $\leftarrow$, conjunction, $\wedge$, disjunction, $\wedge$ and set union, $\cup$. Variables are universally quantified unless otherwise stated. Assume that S is some definition of the state of a computational process and that t(S) denotes a terminating state while $n(S_1, S_2)$ is a computational operation advancing state S_1 to state S_2. Then we can define $c(S_1, S_2)$ to denote that two states, S_1 and S_2, to be related via computation as follows:

$$c(S_1, S_2) \leftarrow (\ t(S_1) \wedge S_2 = S_1) \vee (n\ (S_1, S_n) \wedge c(S_n, S_2))$$

We can define many sequential computations using a refinement of this specification. The primitive operations of the social computer are performed by the entities connected to it that are, potentially, able to collaborate in computations. These will often be humans operating through devices but they could also be automated systems (e.g. 'intelligent' sensors).

The computer requires three different types of data structure:

- A specification, I, of an interaction required by an entity. This can be in any formal language capable of describing the computation needed to coordinate the corresponding interaction.
- A record, S, of the social group currently engaged with I. This can be any appropriate structural representation of the group (the simplest being a set of the names of the entities involved).
- A record, D, of the data associated with each interaction or with each entity. This can be in any appropriate structural representation that allows annotation of the data.

We write Δ to denote the triple (I,S,D) bringing together the three types of data structure. This triple defines the state of a social interaction.

A step in the computation corresponds to a single change in the interaction as a consequence of engagement with it by an entity. We write $m(\Delta_1, \Delta_2)$ to denote an elementary step in the computation of a social interaction. In order to advance from state to state it is necessary for some entity, X, to engage with the interaction, I, in Δ_1 and to perform the elementary computation currently required of it, creating an updated interaction, I_1, and extending the associated data to $D_1 \supseteq D$. We write e(I, X) to denote that X has chosen to engage with interaction I. We write $c(X, I, D, I_1, D_1)$ to denote the computation performed once X is engaged. We then define m as follows:

$$m((I, S, D), (I_1, S \cup \{X\}, D_1)) \leftarrow \exists X.\ e(I, X) \wedge c(X, I, D, I_1, D_1)$$

This abstract definition, as intended, suggests that, but does not define how, certain processes take place. We abstract away from these, because in defining a social machine the entities, X, may be human or artificial.

But the abstraction also places a spotlight on issues which must be resolved if the social machine is to operate. In the case of the definition of m, for example, we leave undecided the *engagement problem* of how X decides to participate in I, and the *alignment problem* of how X knows how to perform the elementary computation, c, required for m. Similarly, a computation in the social computer can be initiated at any time by any entity. We write i(X, Δ) to denote that entity, X, has initiated social interaction Δ. Here we have left undefined the *articulation problem* of how X discovers, infers or invents Δ if X is a human.

These issues – particularly of engagement – impact the question of privacy both for the individual within the social machine and for the machine itself. The sensitivity of information demanded by the machine, the way it will be treated by the machine, and the trust that the individual has in the machine's security infrastructure will all be factors which will help the individual decide whether to participate. With respect to articulation and alignment, the individual may have means to protect his or her privacy, in so far as he or she is sensitive about it, and in that case may try to build protection into the interactions he or she contemplates (for example, using a particular type of device).

A computation is obtained by putting together a sequence of elementary steps in an order permitted by the interaction (defined in Δ) and selected through engagement of entities with the steps. We write $s(\Delta_1, \Delta_2)$ to denote a valid computational sequence from interaction state Δ_1 to state Δ_2, defined recursively as:

$$s(\Delta_1, \Delta_2) \leftarrow (\Delta_1 = \Delta_2) \vee (m(\Delta_1, \Delta_n) \wedge s(\Delta_n, \Delta_2))$$

Interactions' sequences are always initiated by individual entities and then the computation proceeds through engagement with (other) entities individually. There is no central processor with responsibility for making a record of what has happened in the social machine, or with creating the above set of endpoints. This is a key difference from 'closed' social networking systems where the global state of interaction is stored by a third party. This peer-to-peer view also, of course, has effects on the privacy of the group – there is not necessarily any central store of information which creates a security bottleneck, but equally there is no authority which can control or deter information sharing.

Interactions are the only means by which data can be shared, which means that entities accumulate data (other than that which they supply themselves) only through engagement with interactions. We write $\alpha(X, D)$ to denote a data acquisition step by entity, X, to access data set D. We define this below with respect to the points of contact with interactions, where $g(X, \Delta)$ means that X is part of the social group in interaction Δ and $d(\Delta)$ is a function returning the data associated with Δ that X is permitted to access.

$$\alpha(X, d(\Delta)) \leftarrow (i(X, \Delta) \vee \exists \Delta_1 . m(\Delta_1, \Delta)) \wedge g(X, \Delta)$$

This gives us a local view of data by any entity, X, as the set of all data accessible from interactions in which it has participated: $\{Z | \alpha(X, D) \wedge Z \in D\}$.

These data are not likely to be of practical value to X unless there is a consistent system of annotation across interaction sequences (that generated each Δ) such that X can understand each accessed datum reliably. This is an *annotation ontology problem*. We also do not define how $d(\Delta)$ determines which data X is permitted to access from Δ. This is a *social data management problem*. The annotation ontology would be the means of expressing privacy requirements, and maybe individual preferences, relating to specific data items, while social data management mechanisms are the means by which the individual learns to understand and respect the privacy requirements of the machine, and by which the community communicates its requirements to its individual members.

Hence with such a characterisation, we can describe how information must flow around the system. When we consider *how* to implement the abstract characterisation, privacy-related concerns will surface – in the examples in this section, while working out the details of engagement, alignment, articulation, the annotation ontology and social data management. This is not intended as an exhaustive list of privacy-related implementation issues, but merely as indicative of how the abstract conception of the social machine gives a framework within which we can think about the privacy concerns of the machine itself, and of the individual participants within it.

6.5.2 Example Application

The model given in the previous section is, of course, too abstract to be of practical use. Its purpose is to argue that a social computer is novel, even at an abstract level, compared to a conventional computation. However, could it be useful? The same abstract model relates to a range of scenarios that are of interest in application, if we consider how a community might use data and computing technology (such as smartphones) to solve some of its problems, and how we can link this back to the model of the previous section.

There are a number of areas where communities and groups are experimenting in this type of network-enabled cooperation. A common use of social computation is to provide a way of propagating and understanding crime- and nuisance-related information within communities (e.g. Brush et al. 2013). A popular application area in transport is in rapid transmission of travel-related data, where people act as sensors for traffic information and relay this (via analytical tools that filter and amplify the information) across the social group (cf. e.g. Chan and Shaheen 2012). An interesting new area of journalism is the propagation of news not by news organizations but by individuals operating on networks to report events that they have witnessed with these events accumulating credence and detail as others contribute commentary (Meadows 2013; Engesser 2014).

In healthcare, applications of social machines challenge existing systems of centralized provision of healthcare support and information by allowing individuals to set up their own healthcare social groups supported by data owned personally and controlled via local devices (Van Kleek et al. 2013). If we concentrate on healthcare as a specific example, we can map this kind of cooperative interaction onto our schema above.

- **Interaction state [Δ = (I,S,D)]:** Interactions typically involve comparatively sophisticated processes – for example, a protocol for forming a discussion group around a particular medical condition or a care pathway that involves various forms of support or expertise at different stages. The social group will vary in size and diversity for different forms of interaction, from large and open social groups (around dieting, for example) to narrowly focused groups (around critical care pathways). The demands for annotation on the data representation will be extensive because the data shared through interaction will be diverse and there will be great need for alignment of understanding of data across the social group.
- **Initiating interactions [i(X, Δ)]:** Since interactions are complex in this domain it is impractical to expect an individual to directly initiate all but the simplest of interactions. We would expect each interaction to be engineered by a specialist and packaged with sufficient annotation for discovery and re-use to be practical. The devices used by people in healthcare social groups must be able to discover and configure these complex interactions without requiring specialist expertise of the people using those devices (unless that expertise would normally be needed to engage safely with the social interaction).
- **Engaging with interactions [e(I, X)]:** The incentive for engagement with interactions will differ depending on the type of interaction and the social group. For discussion groups the incentive might be the offer of shared data from others in the group, so popularity interactions will depend on volume and quality of high quality data.
- **Performing elementary computations [c(X, I, D, I_1, D_1)]:** The computations required of humans within these social groups typically will involve quite

detailed annotation and accounting for provenance of local health–related data. This raises issues of how individuals can readily express such information through available devices.

- **Belonging to the social group [g(X, Δ)]:** Healthcare social groups may be long lasting so incentives may be needed to stay with the group. This may be through shared experience (via shared data on health views); cooperation (via shared interaction on care pathways) or competition (via comparative data on performance in, for example, dieting).
- **Accessing data [d(Δ)]:** In the social computer all data originates locally with entities on the network so the system of annotation for data must interact with the system of data access to preserve privacy while maintaining incentives through data sharing.

6.6 Privacy

In each case, the analysis of these applications required understanding issues that emerge through the discussion of the d(Δ) stage of accessing data, some of which entailed an understanding of how privacy impacted the relevant social machine. In the case of a crime and policing machine, we would need to protect the identities of witnesses and victims. A transport social machine might make information available concerning the beginning and end points of people's journeys, which they might for obvious reasons want to keep private. The citizen journalist needs to preserve his sources, as well as not falling foul of libel laws, while healthcare data is traditionally extremely secure. Privacy issues will also influence other components of the specification. For instance, the g operator describing participation in a social machine may place constraints upon the member's contribution. For instance, a person who joins a healthcare social machine involving discussion between sufferers of a particular condition may be required to volunteer perhaps sensitive personal data about his problem and experiences, on pain of ejection from the group. The group will thrive if more data is shared, and someone who refuses to share data is essentially free-riding on the group's activity.

6.6.1 The Demands of the Group

In the development of a formal model, issues such as these will need to be followed through in some detail, as the specification takes into account how the flow of data will encourage or inhibit participation in the group, and facilitate the successful completion of its intended task. If we take the example of crime and policing, a number of issues concerning privacy emerge at the design stage.

1. The social machine, as noted above, might use open crime data as well as user-generated data. In the UK, open crime data released by the government is anonymised in space (each incident is snapped to a grid point covering a minimum of 8 postal addresses) and time (the data is subject to aggregated release, a month at a time) for privacy reasons. The question for the social machine is whether anonymised data is sufficiently accurate to help its function. For instance, UK open crime data does not allow inferences to be drawn from the data about whether, say, a particular road is more dangerous after dark, or whether some particular demographic group is more at risk in a particular area (which are precisely the sorts of valuable inference that this sort of system can produce). Hence the social machine is able to provide a perspective and a context on the debate between privacy and utility in open data, which is currently somewhat unanchored and dominated by generic data protection issues.
2. The social machine will need to encourage people to join the group, and initiate and engage with interactions. In general, the more data generated by a group, the richer the picture of the environment that it can produce, and the more effective it is at providing services for its participants. Yet this creates a dilemma. The easiest way to encourage people to join is to lower the barriers to entry – for instance, to allow anonymous reporting of crime. On the other hand, this may prevent the machine from working effectively by failing to prevent spam or other subversive material. If, on the other hand, it is required for the machine to determine or partially determine the provenance of each contribution, then there will have to be some kind of identity management system. That might mean that too many participants will be put off by the need to identify themselves (e.g. it may be too tedious if a password was required), so that in that case the machine would not generate sufficient data to be viable.
3. There will be issues as to who sees the data generated by the system. Should outsiders be able to see it (e.g. the police)? It may be that for the machine to produce effective anti-crime action (as opposed to a discussion forum), that the data should be made available either in the open (for example, superimposed with a map of the community), or sent to the police. On the other hand, it may be that the data could be anonymised (for example, so that the sender was not identifiable).
4. Similarly, there will be issues as to which participants see which data, and whether data is sent to participants on a push or a pull model. Should I have the right to see all the data relevant to the geographical area, or only when I specifically request it, or should my right be restricted to data about the crimes or incidents in which I have an interest?
5. The machine may also require historical data, and so might collect data to provide diachronic analysis of trends. Alternatively, it may delete data from the system once an incident is closed, in which case it will not have access to trend data.

In each case, as argued above, the conditions of information flow will have an effect on information supply, and on the extent of participation in the system. Hence as the design of the social machine unfolds, aspects of the privacy requirements of the successful function of the machine become clear. The machine may require its participants to share data (possibly more than they would prefer), either with each other, or with the outside world. It may require its participants' personal data to be kept strictly within the group, thereby setting out a restrictive privacy principle covering the group itself. It may also require particular privacy policies for other systems with which it interacts.

6.6.2 The Demands of the Individual

As noted earlier, we claim that this sense of group privacy is not derivative from the privacy preferences of the individual participants in the machine. The machine makes demands that may or may not be in accordance with its members' preferences. Of course, those members who object may simply refuse to participate in the machine, thereby removing any inconsistency, but it is conceivable that a member who objected to the machine's privacy requirements may remain as a participant precisely because he gained sufficient extra value from membership of the machine to offset his objection.

A social machine would of course be subject to the same issues of free riding and collective action that we find elsewhere. Perhaps one could imagine social computation being the only or best provider of a specific service, in which case no doubt issues of legitimacy would be raised. It might be argued, for example, that social machines are merely a specific, technology-mediated, instance of the general phenomenon of social control mechanisms to which we are habituated and to which we readily adapt (Schoeman 1992).

On the other hand, one could also very easily imagine a social machine being the locus of a politics of resistance, where people work out their own means of protecting privacy within the machine. In the general peer-to-peer specification we have imagined above, there is no central authority against which one could argue, but certainly some social machines would be centralised (i.e. its specification would not use the peer-to-peer elements of our general specification language). In that case, there may be a power struggle. But one could also use minor resistance strategies, such as the use of what has been called social steganography (boyd and Marwick 2011), or telling trivial lies (Van Kleek et al. 2015).

It is the hope of researchers in this field (O'Hara et al. 2013) that social machines could be driven by participants in a genuine peer-to-peer structure, so that centralised authority will be less prominent and individuals will be empowered to change governing structures, rules and norms. This would, in effect, mean the 'program' of a social machine being changed from within by the machine's components (which is why a social machine is not a Turing machine in the classic sense), or put another way, the machine would be co-opted by a group of participants and its goals and procedures altered. This action would involve understanding how the specification is being rewritten from the bottom up, and would require a new understanding of what the new goal of the machine might be. Naturally enough, the privacy requirements of this new machine, or new phase of the old machine (however it should be conceptualised), may also change, and it is not unlikely that a change in a machine might be driven by the privacy preferences of its individual participants (so, for instance, the goal of a machine may become less ambitious because people are less willing to share or publish information than the original designers expected).

Equally, if the technology takes off and becomes routinely incorporated into governance and service provision, then it is possible that social machines could become instruments of social control. Privacy decisions might once more, as with social networking sites, be taken out of people's hands.

Our point in this paper is not to predict the future. We are neither utopian nor dystopian, though the potential of the technology invites both reactions. Our claim is merely that the function of social machines can be specified in ways that allow us to reason about individual and group privacy. We can begin to answer questions like: what would happen if we stopped publishing this information? Can this function be performed without personal data being shared with the machine? Are the machine's current arrangements optimal with respect to its goal? What is the value of storing the data for a period of time? Social machine theory certainly cannot answer such questions, but it will enable us to determine what the consequences of particular approaches might be.

6.7 Conclusion

In this paper, we have considered trends in technology, particularly associated with the democratisation of use of big data and broad data, as a means for contextualising and reasoning about a non-derivative notion of group privacy. The demands of a social machine on the personal data of its members may, on occasion, strongly affect its function, and hence the integrity of a social machine might create privacy requirements which are only indirectly related to the privacy preferences of its participants. Those requirements may be uncovered by a formal analysis or design of a machine, which, though it will abstract away from social and psychological factors (because such an abstraction is intended to be neutral over the question of whether some particular computation or interaction is carried out by a human, a collective or an artificial agent).

Of course, this method for specifying group privacy requirements only covers groups where people, networked by technology, are consumers of data in pursuit of particular goals, and this may be only a minority pursuit. It certainly does not cover some of the most egregious problems which have driven the immediate academic study of group privacy, such as profiling, surveillance and targeted marketing. Hence it cannot claim to be a universal account of group privacy.

Because it focuses on the integrity, coherence and effectiveness of the group, rather than on the preferences of its participants, this method also separates the specification and description of group privacy from the normative and moral issues it raises. Only when a group has been specified, and its privacy requirements understood, do we then move on to ask whether the social machine is a good or a bad thing. Maybe it is a legitimate social aim to disrupt the social machine – for example, a coalition of cybercriminals linked by remote networked technology is likely to have very firm privacy requirements (Lusthaus 2012). It will need a sophisticated trust management system, and a means for ensuring that the police are unable to infiltrate the machine if its members only communicate remotely. That creates privacy requirements for the machine to function, but equally one would imagine that society in general would support the police in their efforts to disrupt the machine, rather than protect the cybercriminals' privacy rights.

The computational stance we have taken in this paper affords at least two advances in thinking about the group privacy issues related to this type of technology-mediated community. First, the salience of group privacy has been raised specifically by the big data revolution (Floridi 2014), and social machines are consummate creators and consumers of big data. Hence they provide an important and non-artificial context for the debate. Second, in that context privacy can be linked to something tangible – the functionality of the social machine. We can talk about, maybe even quantify, the effects of too little or too much privacy protection before we start to wrestle with the knotty problems of preferences, harms and rights. To repeat, that is not to argue that social machines must always succeed, but that we can specify in a computational language what needs to happen for a social machine to succeed, and then we can move onto the moral debate about whether its success is acceptable or justified (questions which of course cannot be addressed within the computational paradigm).

The integrity of social machines, then, is potentially a way of gaining a non-derivative idea of group privacy, closely linked with the use of online technology and big and broad data to achieve goals for groups and communities, abstracted away from moral and ethical issues concerning whether or not, and when, privacy is a good thing.

Acknowledgements This research was supported by the EPSRC project SOCIAM: The Theory and Practice of Social Machines, ref EP/J017728/1.

Bibliography

Allen, A. L. 2003. Privacy isn't everything: Accountability as a personal and social good, *Alabama Law Review*, 54.

Berners-Lee, T. 1999. *Weaving the web: The original design and ultimate destiny of the world wide web*. New York: HarperCollins.

Bernstein, A., M. Klein, and T.W. Malone. 2012. Programming the global brain. *Communications of the ACM* 55(5): 41–43.

Boyd, D, and A. Marwick. 2011. *Social steganography: Privacy in networked publics*. http://www.danah.org/papers/2011/Steganography-ICAVersion.pdf.

Brush, A. J. B., J. Jung, R. Mahajan, and F. Martinez. 2013. Digital neighborhood watch: Investigating the sharing of camera data amongst neighbors. In *Proceedings of the 2013 conference on computer supported cooperative work*, 693–700. New York: ACM Library.

Chan, N.D., and S.A. Shaheen. 2012. Ridesharing in North America: Past, present and future. *Transport Reviews* 32(1): 93–112.

De Roure, D. 2014. The emerging paradigm of social machines. In *Digital enlightenment yearbook 2014: Social networks and social machines, surveillance and empowerment*, ed. K. O'Hara, C.M. Nguyen, and P. Haynes, 227–234. Amsterdam: IOS Press.

Engesser, S. 2014. Towards a classification of participatory news websites: Comparing heuristic and empirical types. *Digital Journalism* 2(4): 575–595.

Floridi, L. 2014. Open data, data protection, and group privacy. *Philosophy of Technology* 27: 1–3.

Hendler, J., and T. Berners-Lee. 2010. From Semantic Web to social machines: A research challenge for AI on the world wide web. *Artificial Intelligence* 174(2): 156–161.

Hildebrandt, M. 2012. The dawn of a critical transparency right for the profiling era. In *Digital enlightenment yearbook 2012*, ed. J. Bus, M. Crompton, M. Hildebrandt, and G. Metakides, 41–56. Amsterdam: IOS Press.

Jennings, N.R., L. Moreau, D. Nicholson, S. Ramchurn, S. Roberts, T. Rodden, and A. Rogers. 2014. Human-agent collectives. *Communications of the ACM* 57(12): 80–88.

Kleinberg, J., and P. Raghavan. 2005. Query incentive networks. In *Proceedings of the 46th annual IEEE symposium of Foundations of Computer Science (FOCS'05)*, 132–141. Pittsburgh.

Kwiatkowska, M., R. Milner and Vladimiro Sassone. 2004. Science for global ubiquitous computing. *Bulletin of the European Association of Theoretical Computer Science*, 82, 325–333, http://eatcs.org/images/bulletin/beatcs82.pdf.

Lintott, C.J., K. Schawinski, A. Slosar, K. Land, S. Bamford, D. Thomas, M. Jordan Raddick, R.C. Nichol, A. Szalay, D. Andreescu, P. Murray, and J. Vandenberg. 2008. Galaxy Zoo: Morphologies derived from visual inspection of galaxies from the sloan digital sky survey. *Monthly Notices of the Royal Astronomical Society* 389(3): 1179–1189.

Lusthaus, J. 2012. Trust in the world of cybercrime. *Global Crime* 13(2): 71–94.

MacKinnon, C.A. 1989. *Toward a feminist theory of the state*. Cambridge, MA: Harvard University Press.

Meadows, M. 2013. Putting the citizen back into journalism. *Journalism* 14(1): 43–60.

Mill, J. Stuart. 1859. *On liberty*. London: John W. Parker & Son.

Morrow, N., N. Mock, A. Papendieck, and N. Kocmich. 2011. *Independent eEvaluation of the Ushahidi Haiti project*, Development Information Systems International, http://ggs684.pbworks.com/w/file/fetch/60819963/1282.pdf.

O'Hara, K. 2011. *Transparent government, not transparent citizens: A report for the cCabinet office*, London: Cabinet Office, https://www.gov.uk/government/publications/independent-transparency-and-privacy-review.

O'Hara, K. 2012. Trust in social machines: The challenges. In *Proceedings of the AISB/IACAP world congress 2012: Social computing, social cognition, social networks and multiagent systems (SOCIAL TURN/SNAMAS)*, http://eprints.soton.ac.uk/339703/.

O'Hara, K., N.S. Contractor, W. Hall, J.A. Hendler, and N. Shadbolt. 2013. Web science: Understanding the emergence of macro-level features on the world wide web. *Foundations and Trends in Web Science* 4(2/3): 103–267.

O'Hara, K., M.H.C. Nguyen, and P. Haynes. 2014. Introduction. In *Digital enlightenment yearbook 2014: Social networks and social achines, surveillance and empowerment*, ed. K. O'Hara, M.H.C. Nguyen, and P. Haynes, 3–21. Amsterdam: IOS Press.

Okolloh, O. 2009. Ushahidi, or "testimony": Web 2.0 tools for crowdsourcing crisis information. *Participatory Learning and Action* 59(1): 65–70.

Pickard, G., Iyad Rahwan, Wei Pan, Manuel Cebrian, Riley Crane, Anmol Madan and Alex Pentland. 2010. *Time critical social mobilization: The DARPA network challenge winning strategy*, arXiv.org 1008.3172v1, http://hd.media.mit.edu/tech-reports/TR-660.pdf.

Robertson, D., and F. Giunchiglia. 2013. Programming the social computer. *Philosophical Transactions of the Royal Society A: Mathematical, Physical and Engineering Sciences* 371(1987).

Robertson, D., L. Moreau, D. Murray-Rust, and K. O'Hara. 2014. An open system for social computation. In *Digital enlightenment yearbook 2014: Social networks and social machines, surveillance and empowerment*, ed. K. O'Hara, M.H.C. Nguyen, and P. Haynes, 235–252. Amsterdam: IOS Press.

Rosenblum, N.L. 2000. *Membership and morals: The personal uses of pluralism in America*. Princeton: Princeton University Press.

Rössler, B. 2005. *The value of privacy*. Cambridge: Policy Press.

Schoeman, F.D. 1984. Privacy: Philosophical dimensions of the literature. In *Philosophical dimensions of privacy: An anthology*, ed. F.D. FSchoeman, 1–33. Cambridge: Cambridge University Press.

Schoeman, F.D. 1992. *Privacy and social freedom*. Cambridge: Cambridge University Press.

Seiler, M. J., A. J. Collins, and N. H. Fefferman. 2011. *Strategic default in the context of a social network: An epidemiological approach*, Research Institute for Housing America, http://www.housingamerica.org/RIHA/RIHA/Publications/78456_10923_Research_RIHA_Default_Report.pdf.

Shadbolt, N., D. Smith, E. Simperl, M. Van Kleek, Y. Yang, and W. Hall. 2013. Towards a classification framework for social machines. In *Proceedings of SOCM2013: The theory and practice of social machines*, Rio, http://eprints.soton.ac.uk/350513/.

Smith, A. 1994. *An enquiry into the nature and causes of the wealth of nations*, 2 volumes, Indianapolis: Liberty Fund.

Solove, D.J. 2006. A taxonomy of privacy. *University of Pennsylvania Law Review* 154(3): 477–560.

Van Kleek, M., D. Smith, W. Hall, and N. Shadbolt. 2013. "The crowd keeps me in shape": Social psychology and the present and future of health social machines. In *Proceedings of SOCM2013: The theory and practice of social machines*, Rio, http://eprints.soton.ac.uk/350511/.

Van Kleek, M., D. Murray-Rust, A. Guy, D. Smith, and N. Shadbolt. 2015. Self curation, social partitioning, escaping from prejudice and harassment: The many dimensions of lying online. In *2015 ACM web science conference*, Oxford.

von Ahn, L., M. Blum, N.J. Hopper, and J. Langford. 2003. CAPTCHA: Using hard AI problems for security. In *Advances in cryptology: EUROCRYPT 2003*, ed. E. Biham, 294–311. Berlin: Springer.

von Ahn, L., B. Maurer, C. McMillen, D. Abraham, and M. Blum. 2008. reCAPTCHA: Human-based character recognition via web security measures. *Science* 321: 1465–1468.

Walker, P. 2013. Boston bombing identification attempts on social media end in farce. *The Guardian*, 19th April, 2013, http://www.guardian.co.uk/world/2013/apr/19/boston-bombing-suspects-reddit-social-media.

Zinnbauer, D. 2014. *Crowd-sourcing corruption: What petrified forests, street music, bath towels and the Taxman can tell us about the prospects for the future*, Social Science Research Network, http://papers.ssrn.com/sol3/papers.cfm?abstract_id=2508606.

Chapter 7
Indiscriminate Bulk Data Interception and Group Privacy: Do Human Rights Organisations Retaliate Through Strategic Litigation?

Quirine Eijkman

Abstract Human rights groups are increasingly calling for the protection of their right to privacy in relation to the bulk surveillance and interception of their personal communications. Some are advocating through strategic litigation. This advocacy tool is often chosen when there is weak political or public support for an issue. Nonetheless, as a strategy it remains a question if a lawsuit is strategic in the context of establishing accountability for indiscriminate bulk data interception. The chapter concludes that from a legal perspective the effect of the decision to litigate on the basis of the claim that a collective right to group privacy was violated has not (yet) resulted in significant change. Yet the case study, the British case of human rights groups versus the intelligence agencies, does seem to suggest that they have been able to create more public awareness about mass surveillance and interception programs and its side-effects

Keywords Communications surveillance • Bulk data interception • Strategic litigation • Cyber intelligence agencies • Human rights groups • Accountability

Human rights organisations are calling for the protection of group privacy in relation to the indiscriminate bulk interception of their internet and telecom communications. On the basis of the Snowden revelations organisations claim that their personal communications have been collected, inspected, mined, retained and shared by cyber intelligence and/or security agencies. Because until now, they feel that there has been modest political accountability for mass surveillance programs, some human rights organisations are advocating on a group privacy platform through strategic litigation. Strategic litigation concerns the identifying and pursuing of social justice challenges in legal proceedings that may affect changes in law

Q. Eijkman (✉)
HU University of Applied Sciences, Utrecht, Netherlands
e-mail: quirine.eijkman@hu.nl

L. Taylor et al. (eds.), *Group Privacy*, Philosophical Studies Series 126,
DOI 10.1007/978-3-319-46608-8_7

or policy and establishes legal precedents. This advocacy tool is often chosen when there is weak political or public support for an issue. Simultaneously strategic litigation seeks to influence political and public opinion (Amon et al. 2015; Barber 2012; InterRights 2015).[1]

Since Snowden leaked classified cyber intelligence documents in the United Kingdom (UK), France, Germany, the Netherlands and the United States of America (US), human rights organisations are trying to hold intelligence and/or security agencies accountable for their mass surveillance programs (Euro Parliament 2013; Free Snowden 2015). Some focus on the privacy of ordinary people, whereas others are concerned about the privacy of specific groups such as non-governmental organisations (NGOs) or journalists. For instance, Amnesty International staff's communications have been intercepted, whereas the Bureau of Investigative Journalism is concerned about the lack of protection of journalists (Investigatory Powers Tribunal [IPT] 2015c; McLaughlin 2015; Bureau of Investigative Journalism 2014). Strategic litigation challenges how communications interception by cyber intelligence or security agencies affects the grouping of social justice organisations. From a privacy perspective, they argue that activists are not just individual targets, but, because of their work, they have been targeted as a group. Their claims to group privacy are rooted in human rights law, which recognises the need for special safeguards for particular groups such as journalists. Although human rights organisations do not seem to have a similar status, they argue that special safeguards against mass surveillance and communications interception are necessary. Their role and in advocating social justice in a democratic society should, just like journalists, be recognised and protected. They base their claim on international human rights law (Privacy International 2015b, para. 68). For example, articles 8(2) and 10(2) European Convention on Human Rights ('the Convention') requires that states should have effective safeguards against arbitrary interference. An effective safeguard is, for example, prior judicial authorisation (ECHR 2010).

Despite the fact that NGOs are not the only group at risk and strategic litigation is pursued by others[2] and in relation to broader mass surveillance concerns,[3] in this chapter the case study of several human rights organisations: Liberty (The National Council of Liberties), Privacy International, the American Civil Liberties Union & others[4] (ACLU), Amnesty International Limited and Bytes For All against the British Government Communication Headquarters (GCHQ), the Security Service

[1] In comparison to the United States of America (US) strategic litigation or public interest litigation in Europe is a relatively recent phenomenon (Chichowski 2007).

[2] In the UK, for example, two parliamentary representatives of the Green Party have filed legal complaints (Guardian 2014).

[3] The Dutch Citizens versus Plasterk case, for example, concentrates on international bulk data sharing between the Dutch Intelligences and Secrete Services (AIVD and MIVD) and the NSA. A group of human rights organisations, professional organisations and citizens had filed the complaint (*Citizens v. Plasterk* 2014). The case is now in the appeal phase.

[4] These others are the Canadian Civil Liberties Association, the Egyptian Initiative for Personal Right, the Hungarian Civil Liberties Union, the Irish Council for Civil Liberties, the Legal Resources Centre.

& Others and the Secretary of State for the Foreign and Commonwealth Office & Others, is selected (hereafter *NGOs v. Intelligence Services*) (IPT 2014/2015a/b/c). The legal challenge of the *NGOs v. Intelligence Services* case was chosen, because it 'partly' focuses on the right to privacy of a particular group: human rights activists and organisations.[5] Furthermore, strategic litigation is just one advocacy tool of NGOs to challenge alleged indiscriminate bulk data interception. In other words, the organisations involved are not specialised in public interest litigation, such as the former International Centre for the Legal Protection of Human Rights (InterRights), or primarily advocate on a mass surveillance platform. Additionally, in the *NGOs v. Intelligence Services* case there was a judgment by a legal entity, the Investigatory Powers Tribunal, which is now at the European Court of Human Rights (the 'Court') (Amnesty International 2015a; Privacy International 2015b).[6]

Last but not least, the legal challenge of *NGOs v. Intelligence Agencies* reflects the paradox that mass surveillance creates groups, but also hinders them from invoking their rights as a collective. Communications' interception leads to the grouping of personal data, which may be subjected to further investigation. Yet the affected persons can only invoke this right individually and not collectively. Henceforth mass surveillance leads to the creation of groups, but the affected people can only invoke their individual right to privacy. In other words, cyber-intelligence and security agencies such as the American National Security Agency (NSA) or the British GCHC supposedly engage in indiscriminate bulk data interception, which affects groups of people or entities, but NGOs as a collective entity cannot hold them accountable.

Nonetheless, as a strategy it remains a question whether or not an advocacy tool such as a lawsuit is strategic in the context of indiscriminate bulk data interception? In particular addressing the issue of protecting group privacy? What is the goal of human rights organisations in a legal case like the *NGOs v. Intelligence Agencies*? Creating political accountability or public awareness about 'new' human rights violations: indiscriminate bulk data interception conducted by cyber intelligence agencies? Establishing social change through the ECHR (Hodson 2013)? Mobilizing the public on an anti-mass surveillance platform? Or, do they seek to enforce, change, clarify or create new law through jurisprudence that recognizes a new phenomenon: natural persons who protect the rights of groups (Public Law Project 2014)? Is it strategic to focus on a collective privacy violation? And if so, what is the basis of their privacy claim: their groupness? Or did they primarily join forces to put the most weight behind their suit? Last but not least, how can the effect of strategic litigation on the protection of group privacy be determined? The effect is defined as the

---

[5] Many other legal challenges of mass communications surveillance programs focus on others elements such as the bulk sharing of data, hacking, unreasonable search and seizure or freedom of expression.

[6] Other human rights organisations including Big Brother Watch have immediately, without sending a complaint to the Investigatory Powers Tribunal, filled an application to the European Court of Human Rights (ECHR 2014), which is currently being considered.

change brought about by the strategic litigation initiative on communications surveillance and interception programmes or practices.

This chapter focuses on the effect of strategic litigation as an advocacy tool in relation to indiscriminate bulk data interception by intelligence and security agencies. After briefly introducing mass surveillance programs and (bulk) interception practices, a recent case study that partly entailed a group privacy claim is considered. This is followed by a discussion of this tactical choice. Finally, the conclusion reflects on whether or not public litigation on the basis of group privacy is a strategic advocacy tool for human rights organisations in relation to accountability for indiscriminate bulk data interception.

7.1 Mass Surveillance and Indiscriminate Bulk Data Interception

The 2013 Snowden leaks have led to fierce social advocacy, which challenges the alleged indiscriminate bulk data interception of cyber intelligence agencies across the globe. Although mass surveillance and indiscriminate bulk data interception primarily aims to deter or anticipate to national security threats, the programs of among others the American NSA and the British GCHQ that collect data of specific targets or metadata sort side-effects. Snowden's revelations of surveillance and interception programs including PRISM (US-984XN), Upsteam, Boundless Informant, Optic Nerve and the data base X-Keyscore has led to more awareness about the interception of the personal data of individuals and groups. Cyber intelligence and/or security agencies collect this data, while the vast majority of these people do not pose a direct threat.

In public discourse these aforementioned cyber agencies' practices are labelled as 'mass surveillance' (also known as dragnet or blanket surveillance), thereby emphasising that "the garnering of personal data for detailed analysis" entails "in essence the systematic monitoring of large numbers of people without discriminants" (Lyon 2003, p.1). Overall the goals of investigation or monitoring by intelligence and security agencies is to identify individuals of particular interest, but in order to do so a far larger group is affected. In other words the digital personal data of an indiscriminate number of internet or telecom users are collected, inspected, mined, retained and shared. This bulk interception, either through fiber-optic cables, satellites, private actor cooperation, or other means differs from more focused 'targeted' interception, which entails the investigation of a particular entity or person and their networks, who may pose a threat – for example, if someone is potentially involved in cyber espionage.

Since the Global War on Terror the debate on the balance between communications surveillance and privacy has re-emerged. Although this development has been influenced by the increased global reliance on innovative security technology, Snowden's revelations have led to more awareness about alleged indiscriminate bulk data interception and its effect on privacy. Unprecedented opportunities for interception and information-sharing across cyberspace have been created. The digital personal data from, for example, cookies (digital traces of visiting a website) or mobile phone apps are collected and used for pre-emptive threat analysis. For instance, in 2012 personal information was routinely stored by NSA and GCHQ from the game applications of Angry Birds, which were installed on leaky Android smartphones and tablets (Glanz et al. 2014). Even though intelligence and security agencies and increasingly also private actors have always collected personal data to protect national security, databases no longer store information on the selected few, but focus on a far wider pool of potential targets and networks (Amoore 2014). Hence, while protecting society from terrorist attacks is legitimate and the use of interception to achieve this goal may be necessary in a democratic society, the apparent scope (proportionality) of the revealed cyber intelligence surveillance and interception programs is unprecedented.

As Richard Clark (2014), a former presidential cyber security advisor and one of the authors of the American government's report reviewing the data collection and monitoring capabilities at the NSA, stated, "we have created the potential of a police surveillance state". He thereby suggests that even though according to domestic law the indiscriminate bulk data interception was legal, there are questions concerning the proportionality. The means employed to collect digital data should reasonably justify the aim: dealing with the actual national security threat. On the basis of the in 2013 International Principles on the Application of Human Rights to Communication Surveillance[7] the necessity and proportionality threshold should also consider the sensitivity of the personal data and the severity on the infringement of privacy. Also less invasive techniques should be considered, excessive data must be deleted and access to the information should be limited to the agency and purpose for which an authorisation has been provided. For example, the British Tempora program probably did not meet these aforementioned criteria. Central to this 'alleged' surveillance program is the interception and retention of bulk internet traffic data from undersea internet cables. Although in theory it was directed at external communications, between non-UK residents or between a UK resident and a non-resident, in practice it affected a considerable number of users of transatlantic cables: the UK is a key landing point for fibre-optic cables.

Since 2008 the Tempora program has enabled GCHQ to tap into these cables, which carry 10 gigabytes of data per second, and share the collected information with the NSA. Fully operational in 2012, when over 200 interceptions were placed on the fibre-optic cables located off the South West coast of the UK, it did not differentiate between selected and unselected targets. Thus the scope of collecting all internet users' data appears to have been indiscriminate. Furthermore, valuable content data could be kept for 3 days and metadata up to 30 days (Amnesty International 2014; Euro Parliament 2013; Expert Witness Statement 2013; Der Spiegel 2014; MacAskill et al. 2013; The Guardian 2013a). It is unclear whether or not less invasive techniques had been considered, but access to the information was not limited

[7] *International Principles on the Application of Human Rights to Communications Surveillance* (10 July 2013). Available at: https://en.necessaryandproportionate.org/text

to British intelligence and security agencies. Information collected though the Tempora program was probably shared with the NSA (Greenwald and MacAskill 2013). Thus, since the Snowden revelations questions have been raised concerning the scale of privacy infringement by surveillance and interception programs such Tempora.

As such communications surveillance programs and interception practices do not interfere with the right to privacy as enshrined in international and domestic laws including Article 12 Universal Declaration of Human Rights (UNHR), Article 17 of the International Covenant on Civil and Political Rights (ICCPR) or Article 8 of the Convention. However the interference should be based on a specific law, have a legitimate aim, be necessary and proportionate. Also, surveillance programs and interception practices should be subjected to independent and effective oversight (UN General Assembly 2013; UN High Commissioner for Human Rights 2014). As was highlighted in a report of the UN High Commissioner for Human Rights (2014), accountability mechanisms such as – public – warrants provided by a judicial authority, expert oversight, parliamentary oversight committee, administrative review and/or internal procedures are often considered to be adequate safeguards (Wills 2007). In addition, a free press and whistle-blowers may provide for some extra external accountability (Buckland and Wills 2013).

7.2 The Non-governmental Organisations Versus the British Intelligence and Security Services Case

In the *NGOs v. Intelligence Services* case several human rights organisations challenge the 'alleged' indiscriminate communications interception under the Tempora program and information-sharing with the US (including receipt of communications obtained through the surveillance and interception programs PRISM and Upstream). The NGOs main claim is that particular interception activities of GCHQ, MI5 and MI6 violated their human rights under the Convention. Because this alleged indiscriminate interception of personal communications is believed to be neither 'in accordance with the law' nor a 'proportionate interference' with their right to privacy (Article 8) and freedom of expression (Article 10), they argue that their work as human rights organisations and activists has been affected (para.3–6, IPT 2014; COE 2015; Harding 2014). For the involved NGOs such as the Pakistani Bytes For All, it means that they as non-UK human rights activists are concerned about the limited safeguards for non-UK residents. To cite their director Shahzad Ahmad:

> *.....the UK intercept(s) communications in and out of the UK on a mass scale, but it can provide those private communications to foreign governments – including Pakistan – with few restrictions. The idea that the UK is not obliged to offer any privacy protections or safeguards to individuals outside of Britain when conducting surveillance is absurd, and puts at risk the privacy and free expression of human rights activists around the world* (Bytes for All 2014).

In other words they wonder about the impact of indiscriminate bulk data interception on human rights activism across the globe is: what information has GCHQ intercepted, shared and with whom?

Although the human rights organisations in the *NGOs v. Intelligence Services* case and in similar strategic litigation efforts have submitted their claim to the Investigatory Powers Tribunal, this was certainly not a self-evident decision. The somewhat secretive Investigatory Powers Tribunal is an intelligence oversight mechanism that investigates and determines among other things the eligibility of covert activities of the British intelligences agencies (article 65 RIPA; ISC, 2015). Their credibility is questioned by human rights organisations that want to hold cyber intelligence and security agencies accountable for mass surveillance and interception programs. In 2014, for instance, in the *Big Brother and Others v. the UK* case the complainants immediately applied to the ECHR. From their perspective there is no effective remedy in the UK (ECHR 2014). One of the challenges for the civil society organisations is that their case is based on 'assumed facts'. The British government has a 'neither confirmed nor denied' existence policy. In contrast to the US that publicly acknowledged the existence of the PRISM and UPSTREAM program, the UK has not publicly admitted or disclosed the Tempora program. Subsequently, human rights organisations do not have to prove that the surveillance programs exist or that their communications have been intercepted unlawfully, but for the majority of involved NGOs it is hypothetical that this may have occurred (para.4 IPT 2015a).[8] Henceforth, even if the claim of human rights organisations in the *NGOs v. Intelligence Services* would be acknowledged by the Investigatory Powers Tribunal, it is unlikely that on the basis of that judgment the British cyber intelligence interception practises will have changed significantly.

Furthermore, the legal basis of the Tempora program's interception is articles s.1(5)b, s.2(2/7) and s.8(4) of the Regulation of Investigatory Powers Act (RIPA). This act regulates the tapping of communications in the UK and requires a warrant signed by the Secretary of State for the interception of internal communication and for other (external) communications. The warrant can also be accompanied by a certificate from the secretary of state, which authorises more indiscriminate trawling (Part 1, Chapter 1 and articles s.8(1) and s.8(4/5) RIPA; Europarliament 2013). Additionally, the intelligence agencies according to the foreign secretary acted in full compliance with the Human Rights Act and the Intelligence Services Act (Guardian 2013b). When taking a closer look as to how the tribunal dealt with the 'group privacy' claim as put forward by the human rights organisations, it appears that in relation to the interception under the Tempora program its key privacy considerations focused on the legal question of whether or not the certificates, which had been authorised by the secretary of state, had been issued lawfully. And, whether or not, from a British statutory framework and Convention perspective, the safeguards for an individual's privacy were adequate (para.5–6 and 83 IPT 2015a). As

[8] In June 2015 two human rights organisations, the South African Legal Resources Centre and Amnesty International, received notice of the Investigatory Powers Tribunal that they had been lawfully subjected to interception and proportionally intercepted and assessed, but that their data had been retained too long (IPT 2015a/b).

the allegedly untargeted interception of their communications by the Tempora program required a warrant in combination with a certificate,[9] the intercepted communications and subsequent recordings could have been considered lawful, and the access, if not indiscriminate, necessary in the interest of national security (para.79, 83 and 159–160 IPT 2014). Therefore, the supposed interception by the British intelligence services did not, according to the judgment, amount to a violation of privacy or arbitrary conduct. Henceforth the interception was ruled to be lawful and the interference by the cyber intelligence agencies justified. Since then a second judgment in the case determined that intelligence sharing between UK and US was unlawful prior to December 2015. This, because until then the procedures for British access to the information collected through the NSA's PRISM and Upstream programs were secret (IPT 2015a). It was the first time in its 15 years of existence that the Investigatory Powers Tribunal ruled in favor of the British intelligence and security services. Last but not least, in June and July 2015 the Investigatory Powers Tribunal acknowledged that two of the involved human rights organisations the South African Legal Resources Centre and Amnesty International had been lawfully subjected to interception and proportionally intercepted and assessed, but that their data had been retained too long (IPT 2015a/b). This was a breach of article 8 of the Convention.

Furthermore, in the *NGO v. Intelligence Agencies* case human rights organisations also put forward that their right to the freedom of expression had been interfered with. As investigatory NGOs, some of the involved human rights organisations, argued that they as a special group required the same protection. How else can the sometimes confidential information they obtain be adequately protected? Nonetheless, the Investigatory Powers Tribunal ruled on the basis of the same reasoning as article 8, the right to privacy, of the Convention that the human rights organisations' right to freedom of expression had not been violated (para. 12–13, 134–149 and 152 IPT 2014)? Hence their special status as a special group – human rights NGOs – which required protection was not acknowledged. In an open brief submitted to the Investigatory Power Tribunal two of the involved organisations, Privacy International and Liberty, had anticipated that the government would take this position (Privacy International & Liberty 2014, para.34–35). Since the two judgments (IPT 2014/2015a, 2015b, 2015c) some of the human rights organisations involved, Privacy International, Liberty and Amnesty International, have brought the case to the ECHR (Privacy International 2015b).

[9]An interception warrant is either targeted (article s.8(1) RIPA) or untargeted/strategic (article s.8(4) RIPA).

7.3 A Violation of Group Privacy?

Strategic litigation is considered by human rights organisations to be a key advocacy tool to bring communications surveillance programs and interception practices in line with human rights. Through the *NGO v. Intelligence Agencies* case the organisations involved are advocating for more awareness about Snowden's revelations and in particular to clarify and reform current alleged indiscriminate bulk data interception practices. From a group privacy perspective, however, one can question whether the claim that the privacy of human rights organisations as a group was violated is strategic. In other words, was the groupness of the privacy violation really a key concern? Amnesty International UK's press release, which announces NGOs bringing the case to the ECHR, is subtle. Their defence counsel Nick Williams formulates it as

> *This industrial scale mass surveillance makes it increasingly difficult for organisations like Amnesty International to carry out human rights work. It is critical that we are able to seek and receive information of public interest from our confidential sources, free from government intrusion* (AI 2015a).

Hence, he is implicitly making the argument that indiscriminate bulk data interception interferes with international human rights activism. This is because there is a risk that people across the world will no longer share personal communications with human rights NGOs out of fear of repercussions from their own states. Subsequently, in violating the privacy of groups there is a risk that the role of human rights organisations is marginalised.

Perhaps the emphasis on group privacy by the different organisations was an effort to substantiate NGO's indiscriminate interception claim before the Investigatory Powers Tribunal. If human rights organisations are not able to make this legal argument they run the risk that their case will be declared inadmissible or they cannot prove that they suffered (individual) harm. Jurisprudence in similar cases indicates that it is challenging for human rights organisations to substantiate the argument that they as a particular group have been affected by indiscriminate bulk data interception by cyber intelligence agencies. For example, in *Clapper v. Amnesty International US* the Supreme Court ruled that human rights organisations lacked standing[10] and therefore they were denied access to the federal court system, whereas in the *Dutch Citizens versus Plasterk* case the NGOs' claimants had standing but according to the court they could not prove that they or the people they represented suffered individual harm (*Clapper v. Amnesty International* 2013; *Citizens v. Plasterk* 2014).

[10] Basically the question was whether or not a group of international NGOs, labour organisations, lawyers and journalist had standing to sue the US alleging that they were imminently collecting their international communications through surveillance under the 1978 Foreign Intelligence Surveillance (FISA) Act. They challenged the constitutionality of the 2008 Amendments Act, which reformed the 1978 FISA Act.

Jurisprudence of the ECHR confirms that the right of complaint is primarily recognized if the claimants, individuals or a group can demonstrate a personal interest and that they have suffered personal harm.[11] Therefore, it is no surprise that in a similar mass surveillance case, such as the Big Brother Watch and Others case, the ECHR is enquiring whether Big Brother Watch and the others could pass the so-called admissibility test. Can they claim to be a victim of a privacy violation under article 8 of the Convention, which is defined as a right of a natural person to protect his or her interests (ECHR 2014)? As van der Sloot (2014) emphases in relation to large-scale data interception by intelligence agencies or private actors the individual interest criteria should be less restrictive. Even though ECHR jurisprudence is consistent in emphasising that the effect of the intervention should be suffered directly, usually labelled as individual harm, in relation to for instance Big Data this is unrealistic. People are simply not aware which of their personal communications have been collected, inspected, mined, retained or shared in data bases across the world. Therefore the admissibility of groups, who may for no apparent reason be singled out by algorithms, should be considered.

The effect of the group privacy claim in the case study, *NGOs v. Intelligence Agencies*, appears to be modest. As an advocacy tool in relation to communications interception, it has generated some (social) media and public attention, but in the judgment from the Investigatory Powers Tribunal there is no reference to an alleged violation of group privacy of the one or more of the involved human rights organisations (para. 12, 153–154, IPT 2014). Social media coverage and news reports in relation to the case did not appear to communicate this group privacy perspective very specifically. Most simply mentioned that the NGOs had brought the legal challenge and referred to violation of privacy in general.[12]

In terms of law the human rights organisations have, however, been able to contribute to a public debate about the lawfulness of Tempora's (bulk) interception practises and whether or not the British safeguards for communications surveillance are sufficient. For instance, in the second judgment the Investigatory Powers Tribunal determined that intelligence sharing between the NSA and GCHQ prior to the first judgment of the tribunal in December 2014 was unlawful. The reason being that the rules governing GCHQ's access to the American UPSTREAM and PRISM programmes were secret (IPT 2014/2015a, 2015b, 2015c). Some of the involved human rights organisations such as Privacy International and Bytes for All intend to ask the Investigatory Powers Tribunal to immediately delete their internet and telecom communications, which until December 2014 had been collected unlawfully (Privacy International 2015a), thereby once more creating an opportunity to emphasise the

[11] See among others (ECHR 2008).

[12] In a Google news search on the day of and the day after the Investigatory Powers Tribunal's judgment on 5 and 6 December 2014, there are 202 news references mentioning the *NGOs v. Intelligence Agencies* case, and on 5 and 6 February 2015, 321 references (IPT 2014/2015a, b, c). Source Google news search. Keywords 'Investigatory Powers Tribunal', 'civil society' and 'privacy' (accessed 18 March 2015).

effect that mass surveillance has on human rights and the fact that the affected NGOs can only invoke their right to privacy after demonstrating personal harm.

7.4 Reflections

The legal challenge of the *NGOs v. Intelligence Agencies* case study reflects the paradox that mass surveillance on the one hand creates groups, but on the other hand prevents them from invoking their right to privacy collectively. Indiscriminate bulk data interception leads to the grouping of personal data, which may be subjected to further investigation. Yet the affected persons can only invoke this right individually. Thus mass surveillance leads to the creation of groups, but the affected NGOs can only invoke their individual right to privacy and not do so as a collective.

Human rights advocacy through public litigation was a strategic choice for human rights organisations in their pursuit of seeking accountability for indiscriminate bulk data interception of their personal communications. There are several legal cases addressing the surveillance and interception programs that were revealed by Edward Snowden. In the case study selected for this chapter, the *NGOs v. Intelligence Agencies* case, the British bulk data interception program Tempora was questioned. In particular its effect on human rights organisations was put forward in the claim. Obviously the legal brief of the involved human rights organisations also included concerns broader than group privacy. Furthermore, the majority are also engaged in other forms of advocacy against mass surveillance and indiscriminate bulk data interception by cyber intelligence and security agencies.

From an accountability perspective the effect of strategic litigation on protecting the right to group privacy against alleged blanket communications surveillance and indiscriminate bulk data interception appears to be modest. The direct change brought about by the strategic litigation initiative on mass surveillance and bulk data interception is challenging to discern. The sharing of intercepted data between the American and British cyber intelligence and security agencies was deemed lawful in the judgment. To some extent, therefore, the alleged indiscriminate bulk data interception was ruled to be legitimate. Also the human rights organisations were neither in the two judgments of the Investigatory Powers Tribunal nor in the media recognised as a specific group whose privacy required special consideration. The NGOs will, however, pursue this argument in their ECHR application. Nonetheless, from a legal and media perspective the decision to litigate on the basis of the claim that a collective right to group privacy was violated has not (yet) resulted in significant change.

However, in terms of raising awareness about the chilling effect of mass surveillance on human rights activism as well as the functioning of accountability mechanisms in the UK, strategic litigation may have had created some effect. On the one hand, even if the complaints had been accepted by the Investigatory Powers Tribunal, it is unlikely that on the basis of that judgment the British cyber intelligence agencies would have been effectively held accountable. The British government pursues its ‘neither confirmed nor denied’ existence policy and therefore the case was hypothetical. On the other hand, by litigating, NGOs have shaped and constructed human rights, which in this case meant that they brought the need for the protection of group privacy for activists at home and abroad into focus (Hodson 2013). Also they obtained a small but significant legal victory before the Investigatory Powers Tribunal, the intelligence sharing between the UK and the US was ruled partly unlawful. Thus the human rights organisations demonstrated that it is possible to hold intelligence and security services accountable, while at the same time emphasising in among others an open brief that the secrecy surrounding this tribunal is interfering with transparent oversight. Furthermore, domestic strategic litigation paved the way for bringing the case to the ECHR, thereby ensuring that NGOs concerns about the chilling effect of indiscriminate bulk interception of a group's communications data, the lack of proper oversight in the UK and bilateral information-sharing will at the international level potentially lead to some form of retaliation.

Bibliography

Amnesty International (AI). 2013. Amnesty International brings claim against UK over state surveillance [Online] Available from: http://www.amnesty.org/en/news/amnesty-international-brings-claim-against-uk-over-state-surveillance-2013-12-09 [Accessed: 16th Feb 2014].

Amnesty International (AI). 2014. Free speech [Online], Amnesty International United Kingdom website, Available from: http://www.amnesty.org.uk/why-taking-government-court-mass--spying-gchq-nsa-tempora-prism-edward-snowden#.VPXW6885BMt [Accessed: 16th Nov 2014].

Amnesty International (AI). 2015a. Amnesty International takes UK government to the European Court of Human Rights over mass surveillance [Online], Amnesty International United Kingdom website, Available from: http://www.amnesty.org.uk/press-releases [Accessed: 25th Mar 2015].

Amnesty International (AI). 2015b. Amnesty International calls upon David Cameron to launch surveillance enquiry [Online], Amnesty International website, Available from: https://www.amnesty.org/en/latest/news/2015/07/amnesty-international-calls-on-david-cameron-to-launch-surveillance-inquiry/ [Accessed: 12th July 2015].

Amon, J. J., M. Wurth, and M. McLemora. 2015. Evaluating Human Rights Advocacy on Criminal Justice and Sex Work, Health and Human Rights Journal, 17/1/[Online] Available from: http://www.hhrjournal.org/2015/01/29/evaluating-human-rights-advocacy-on-criminal-justice-and-sex-work/ [Accessed: 16th Feb 2015].

Amoore, L. 2014. Security and the claim to privacy. *International Political Sociology* 8: 108–112.

Barber, C.C. 2012. Tackling the evaluation challenge in human rights: Assessing the impact of strategic litigation organisations. *The International Journal of Human Rights* 16(3): 411–435.

Bergen, P., D. Sterman, E. Schneider, and C.Bailey. 2014. Do NSA's bulk surveillance programs stop terrorists? [Online] Available from: http://www.newamerica.net/publications/policy/do_nsas_bulk_surveillance_programs_stop_terrorists [Accessed: 5th Feb 2014].

Buckland, B. A. and A. Wills. 2013. Whistleblowing in the security sector. In *Protection of whistleblowers* eds. N. Ruzic, B. Medenica. Belgrade: Commissioner for Information of Public Importance and Personal Data Protection. [Online] Available from: https://whistlenetwork.files.wordpress.com/2014/01/bucklandwills-chapter.pdf [Accessed: 15 Feb 2014].

Bureau of Investigative Journalism. 2014. Surveillance State: Bureau Files ECHR Case Challenging UK Government over Surveillance of Journalists Communications [Online] Available from:

http://www.thebureauinvestigates.com/2014/09/14/bureau-files-echr-case-challenging-uk-government-over-surveillance-of-journalists-communications/ [Accessed: 5 Feb 2015].

Bytes For All. 2014. Investigatory powers tribunal finds GCHQ's tempora program in principle legal, 5 December 2015, [Online] Available from: http://content.bytesforall.pk/node/157 [Accessed: 5th Feb 2015].

Chichowski, R. 2007. *The European court and civil society: Litigation, mobilization and governance*. Cambridge: Cambridge University Press.

Citizens versus Plasterk. 2014. ECLI:N:RBDHA:2014:8966, The Hague District Court, no. C09/445237 HA ZA 13–1325, 23 July 2014.

Clapper v. Amnesty International. 2013. *Clapper, Director of national intelligence versus Amnesty International USA et al.* 568, US Supreme Court, 26 February 2013.

Clark, R. 2014. Richard Clarke at RSA Conference: 10 Observations on US Intelligence Gathering. [Online] Available from: http://windowsitpro.com/identity-management/richard-clarke-rsa-conference-10-observations-us-intelligence-gathering. [Accessed: 2 Mar 2014].

Cortright, D. et al. 2011. Friend not foe: Opening spaces for civil society engagement to prevent violent extremism. [Online] Available from: http://www.hscollective.org/wp-content/uploads/2013/09/Friend-not-Foe-2.pdf. [Accessed: 1 Feb 2014].

Council of Europe (COE). 2015. Mass surveillance, committee on legal affairs and human rights, Strasbourg: Council of Europe, Nr. As/Jur (2015)01, 26 January 2015, http://website-pace.net/documents/19838/1085720/20150126-MassSurveillance-EN.pdf/df5aae25-6cfe-450a-92a6-e903af10b7a2 [Accessed: 23 Feb 2015].

Der Spiegel. 2014. The NSA in Germany: Snowden's documents available for download *Spiegel Online International* 18 June 2014 [Online] Available from: http://www.spiegel.de/international/the-germany-file-of-edward-snowden-documents-available-for-download-a-975917.html [Accessed: 19 Feb 2015].

European Court of Human Rights (ECHR). 2006. *Weber versus Germany.* Application number 54934/00, 29 June 2006

European Court of Human Rights (ECHR). 2008. *Liberty and others versus the United Kingdom.* Application number 58243/00, 1 July 2008.

European Court of Human Rights (ECHR). 2010. *Sanoma and others versus the Netherlands.* Application number 38224/03, 14 September 2010.

European Court of Human Rights (ECHR). 2012. *Telegraaf and others versus the Netherlands.* Application number 39315/06, 22 November 2012.

European Court of Human Rights (ECHR). 2014. *Big brother and others versus the United Kingdom.* Application number. 58170/13, 7 January 2014.

European Parliament. 2013. *National programmes for mass surveillance of personal data in EU member states and their compatibility with EU Law*, Directorate for Internal Policies Policy Department Citizen's Rights and Constitutional Affairs, PE.493.032 http://www.europarl.europa.eu/RegData/etudes/etudes/join/2013/493032/IPOL-LIBE_ET(2013)493032_EN.pdf [Accessed: 23rd Feb 2014].

Expert Witness Statement of Ian Brown for Big Brother Watch and Others (Expert Witness Statement). 2013. Re: Large-Scale Internet Surveillance by the UK Application No. 58170/13, European Court of Human Rights.

Farrell, H., and M. Finnemore. 2013. The end of hypocrisy: American foreign policy in the age of leaks. [Online] Available from: http://www.foreignaffairs.com/articles/140155/henry-farrell-and-martha-finnemore/the-end-of-hypocrisy [Accessed: 15th Mar 2014].

Free Snowden. 2015. Impact. *Free Snowden Website* [Online] Available from: https://www.freesnowden.is/impact/ [Accessed: 19th Feb 2015].

Gellman, B., and A. Soltani. 2014. NSA surveillance program reaches "into the past" to retrieve, replay phone calls. *The Washington Post.* [Online] Available from: http://www.washingtonpost.com/world/national-security/nsa-surveillance-program-reaches-into-the-past-to-retrieve-replay-phone-calls/2014/03/18/226d2646-ade9-11e3-a49e-76adc9210f19_story.html. [Accessed: 19th Mar 2014].

Gellman, B., A. Soltani, and A. Peterson. 2013. How we know the NSA had access to internal google and yahoo cloud data. *The Washington Post.* [Online] Available from: http://www.washingtonpost.com/blogs/the-switch/wp/2013/11/04/how-we-know-the-nsa-had-access-to-internal-google-and-yahoo-cloud-data/. [Accessed: 3rd Nov 2014].

Glanz, J., J. Larson, and A. Lehren. 2014. Spy agencies tap data streaming from phone apps. *New York Times* [Online] Available from: http://www.nytimes.com/2014/01/28/world/spy-agencies-scour-phone-apps-for-personal-data.html?_r=0 [Accessed: 3rd Jan 2014].

Gorvin, I. 2012. Producing the evidence that human rights advocacy works: First steps towards systematized evaluation at human rights watch. *Journal of Human Rights Practice* 1(3): 477–487.

Greenwald, G. and W. MacAskill. 2013. NSA prism program taps in to user data of apple, google and others. *The Guardian.* 6 June 2013 [Online] Available from: http://www.theguardian.com/world/2013/jun/06/us-tech-giants-nsa-data. [Accessed: 12th Feb 2014].

Harding, L. 2014. Edward Snowden: US government deliberately snooped on human rights workers. *The Guardian.* 8 April 2014 [Online] Available from: http://www.theguardian.com/world/2014/apr/08/edwards-snowden-us-government-spied-human-rights-workers [Accessed: 5th Nov 2014].

Hodson, L. 2013. Activating the law: Exploring the legal responses of NGOs to gross human rights violations. In *Towards a sociology of human rights*, ed. M. Madsen and G. Verschraegen, 267–283. Oxford: Hart Publishing.

Inkster, N. 2014. The Snowden revelations: Myths and misapprehensions. *Survival: Global Politics and Strategy* 56(1): 51–56.

Intelligence & Security Committee of Parliament (ISP). 2015. *Privacy and security: A modern and transparent legal framework.* 12 March 2015, London: House of Commons.

InterRights. 2015. Our cases. [Online] Available from: http://www.interights.org/our-cases/index.html [Accessed: 19 Feb 2015].

Investigatory Powers Tribunal (IPT). 2013. The respondents open response, UKIPTTrib 13_77H, Case Nos: IPT/13/77H/IPT/13/92/CH, 15 November 2013, London.

Investigatory Powers Tribunal (IPT). 2014. Judgment (First Judgment), [2014] UKIPTTrib 13_77-H, Case Nos: IPT/13/77/H, IPT/13/92/CH, IPT/13/168-173/H, IPT/13/194/CH, IPT/13/204/CH, 5 December 2014, London.

Investigatory Powers Tribunal (IPT). 2015a. Judgment (Second Judgment), [2015] UKIPTTrib 13 77-H, Case Nos: IPT/13/77/H, IPT/13/92/CH, IPT/13/168-173/H, IPT/13/194/CH, IPT/13/204/CH, 6 February 2015, London.

Investigatory Powers Tribunal (IPT). 2015b. Determination, [2015] UKIPTTrib 13 77-H, Case Nos: IPT/13/77/H, IPT/13/92/CH, IPT/13/168-173/H, IPT/13/194/CH, IPT/13/204/CH, 22 June 2015, London.

Investigatory Powers Tribunal (IPT). 2015c. Post and Email, [2015] UKIPTTrib 13-77-H, Case Nos: IPT/13/77/H, IPT/13/168-173/H, IPT/13/194/CH, 2 July 2015, London.

Kessler, G. 2013. James Clapper's least "untruthful statement" to the Senate. *The Washington Post.* [Online] Available from: http://www.washingtonpost.com/blogs/fact-checker/post/james-clappers-least-untruthful-statement-to-the-senate/2013/06/11/e50677a8-d2d8-11e2-a73e-826d299ff459_blog.html. [Accessed: 25 Feb 2014].

Levi, M., and D.S. Wall. 2004. Technologies, security, and privacy in the post-9/11 European information society. *Journal of Law and Society* 31(2): 194–220.

Lyon, D. 2003. *Surveillance as social sorting: Privacy, risk and social digital discrimination.* Cambridge: Polity Press.

MacAskill, E. et al. 2013. GCHQ taps fibre-optic cables for secret access to world's communications. *The Guardian.* [Online] Available from: http://www.theguardian.com/uk/2013/jun/21/gchq-cables-secret-world-communications-nsa [Accessed: 23 Feb 2014].

McLaughlin, J. 2015. British tribunal flip-flops on wrongful surveillance of Amnesty International. *The Intercept.* [Online] Available from: https://firstlook.org/theintercept/2015/07/01/major-reversal-british-tribunal-confirms-surveillance-amnesty-international-violated-rights/ [Accessed: 13 July 2015].

Newman, M. 2014. Mass surveillance regime does not violate human rights law, tribunal rules, Bureau of Investigative Journalism, 5 December 2014 [Online] Available from: http://www.

thebureauinvestigates.com/2014/12/05/mass-surveillance-regime-does-not-breach-human-rights-law-tribunal-rules/ [Accessed: 7 Feb 2015].

Norton-Taylor, R. 2013. UK intelligence chiefs get off scot-free in grilling on NSA leaks. *The Guardian.* [Online] Available from: http://www.theguardian.com/uk-news/defence-and-security-blog/2013/nov/08/uk-intelligence-grilling-nsa-leaks. [Accessed: 5 Feb 2014].

Norton-Taylor, R., and D. Rushe. 2013. Ex-MI6 deputy chief plays down the damage caused by Snowden leaks. *The Guardian.* [Online] Available from: http://www.theguardian.com/world/2013/sep/12/mi6-plays-down-damage-edward-snowden-leaks. [Accessed: 17 Feb 2014].

Privacy International. 2015a. GCHQ-NSA intelligence sharing unlawful, says UK surveillance tribunal [Online] Available from: https://www.privacyinternational.org/?q=node/482 [Accessed 6 Feb 2015].

Privacy International. 2015b. 10 human eights organizations versus the United Kingdom: additional submissions on the facts of complaints. [Online] Available from: https://privacyinternational.org/sites/default/files/HR%20Orgs%20v%20UK.pdf [Accessed 6 July 2015].

Privacy & Civil Liberties Oversight Board. 2014. *Report on the telephone records program conducted under section 215 of the USA Patriot Act and on the operations of the foreign intelligence surveillance court.* Washington: Privacy & Civil Liberties Oversight Board.

Propublica. 2014. Spy agencies probe angry birds and other apps for personal data. [Online] Available from: http://www.propublica.org/article/spy-agencies-probe-angry-birds-and-other-apps-for-personal-data. [Accessed: 27 Feb 2014].

Public Law Project. 2014. *The guide to strategic litigation.* [Online] Available from: http://www.publiclawproject.org.uk/resources/153/guide-to-strategic-litigation [Accessed: 10 Feb 2015].

Rosenberg, G.N. 2007. *The hollow hope: Can courts bring social change.* Chicago: University of Chicago Press.

Savage, C. 2013. NSA said to search content of messages to and from US. *The New York Times.* [Online] Available from: http://www.nytimes.com/2013/08/08/us/broader-sifting-of-data-abroad-is-seen-by-nsa.html?pagewanted=all&_r=0. [Accessed: 5 Feb 2014].

Schmidt, E., and J. Cohen. 2013. *The new digital age: Reshaping the future of people, nations and business.* New York/Toronto: Random House.

Surveille. 2013. *Deliverable D2.3: Paper by local authorities end-users.* [Online] Available from: http://www.surveille.eu/PDFs/D2.3%20Paper%20by%20Local%20Authorities%20End%20Users.pdf. [Accessed: 2nd Mar 2014].

The Guardian. 2013a. GCHQ taps fibre-optic cables for secrete access to world communication. *The Guardian.* 21 June 2013, [Online] Available from: http://www.theguardian.com/uk/2013/jun/21/gchq-cables-secret-world-communications-nsa [Accessed: 15th Feb 2014].

The Guardian. 2013b. The legal loopholes that allow GCHQ to spy on the world. *The Guardian.* 21 June 2013, [Online] Available from: http://www.theguardian.com/uk/2013/jun/21/legal-loopholes-gchq-spy-world [Accessed: 15th Feb 2014].

The Guardian. 2014. Green politicians launch legal challenge of GCHQ surveillance. *The Guardian.* 4 May 2014, [Online] Available from: http://www.theguardian.com/uk-news/2014/may/04/greens-legal-challenge-gchq-surveillance [Accessed: 5th Feb 2014].

Udall, M., and R. Wyden. 2013. *Statement on reports of compliance violations made under NSA collection programs.* [Online] Available from: http://www.markudall.senate.gov/?p=press_release&id=3666. [Accessed: 6 Mar 2014].

UN High Commissioner for Human Rights. 2014. 'The Right to Privacy in the Digital Age', report by the Office of the UN High Commissioner for Human Rights presented to the United Nations Human Rights Council, 30 June 2014. UN DOC A/HRC/27/37.

United Nations (UN) General Assembly. 2013. 'The Right to Privacy in the Digital Age', 18 December 2013, UN DOC A/RES/68/167.

United Nations (UN) Special Rapporteur. 2009. Report by Special Rapporteur Martin Scheinin (28 December 2009), Human Rights Council, UN Doc A/HRC/13/37.

van der Sloot, B. 2014. Privacy in the Post-NSA Era Time for a Fundamental Revision. *The Journal of Intellectual Property, Information Technology and E-Commerce Law* 5(2): 1–11.

van Gulijk, C. 2014. European commission FP7 project: Surveillance: Ethical issues, legal limitations, and efficiency. [Online] Available from: http://www.surveille.eu/# [Accessed: 10th Mar 2014].

Visser, J. 2014 Wat zijn de gevolgen van het Plasterkdebat? (What are the consequences of the Plasterkdebate) [Online] Available from: http://www.volkskrant.nl/vk/nl/2686/Binnenland/article/detail/3595638/2014/02/12/Wat-zijn-de-gevolgen-van-het-Plasterkdebat.dhtml. [Accessed: 3 Mar 2014].

White House. 2014. Remarks of the President on signals review. [Online] Available from: http://www.whitehouse.gov/the-press-office/2014/01/17/remarks-president-review-signals-intelligence. [Accessed: 3 Mar 2014].

White House Office of the Press Secretary. 2010. Summary of the white house review of the December 25, 2009 attempted terrorist attack 7 January 2010. [Online] Available from: http://www.whitehouse.gov/the-press-office/white-house-review-summary-regarding-12252009-attempted-terrorist-attack [Accessed: 25 Feb 2014].

Wills, A. 2007. *Understanding intelligence oversight.* Geneva: Geneva Centre for Democratic Control of Armed Forces (DCAF).

Chapter 8
From Group Privacy to Collective Privacy: Towards a New Dimension of Privacy and Data Protection in the Big Data Era

Alessandro Mantelero

Abstract This chapter focuses on big data analytics and, in this context, investigates the opportunity to consider informational privacy and data protection as collective rights. From this perspective, privacy and data protection are not interpreted as referring to a given individual, but as common to the individuals that are grouped into various categories by data gatherers.

The peculiar nature of the groups generated by big data analytics requires an approach that cannot be exclusively based on individual rights. The new scale of data collection entails the recognition of a new layer, represented by groups' need for the safeguard of their collective privacy and data protection rights.

This dimension requires a specific regulatory framework, which should be mainly focused on the legal representation of these collective interests, on the provision of a mandatory multiple-impact assessment of the use of big data analytics and on the role played by data protection authorities.

Keywords Big data • Right to privacy • Data protection • Group privacy • Risk assessment

8.1 Introduction

Since their origins, both informational privacy and data protection have been protected as individual rights, even though the social dimension of these rights has been acknowledged and taken into account by courts and data protection authorities, as well as by policy makers. Nevertheless, the rights holder has always been the data subject and the rights regarding informational privacy have been mainly exercised by single individuals.

This approach based on individual rights is consistent with the traditional notion of groups as the sum of the relationships existing among their members. From this perspective, group privacy concerns the peculiar nature of the sharing of personal

A. Mantelero (✉)
Politecnico di Torino, Torino, Italy
e-mail: alessandro.mantelero@polito.it

L. Taylor et al. (eds.), *Group Privacy*, Philosophical Studies Series 126,
DOI 10.1007/978-3-319-46608-8_8

information that takes place within a group. For this reason, it is a sort of context-related notion of individual privacy.[1] However, this atomistic view seems to be inadequate in the existing context of predictive analytics.

In the big data era, new technologies and powerful analytics make it possible to collect and analyse large amounts of data, in order to identify patterns in groups' behaviour.[2] The new element of this group analysis is given by the fact that groups are designed by data gatherers, by selecting specific clusters of information. Data gatherers shape the population they intend to investigate and collect information about different people who do not know the other members of the group and, in many cases, are not aware of the consequences of being part of a group.[3]

The different nature of these groups requires a different approach that cannot be exclusively based on individual rights. The new scale of data collection entails the recognition of another layer, represented by the rights of groups to the protection of their collective dimension of privacy and data.

As in other cases of collective and diffuse interests, attention is due to the nature of the interests that should be protected and to legal remedies. From this perspective, the potential role played by bodies representative of these collective interests should also be considered.

8.2 From Group Privacy to Collective Privacy

The modern notion of the right to privacy draws its origin from the theories elaborated at the end of the nineteenth century, both in the U.S. and Europe.

In the U.S., Warren and Brandeis (1890) shaped the modern idea of privacy, which was different from the previous notion of protection of private life based on property (see Warren and Brandeis (1890); Westin (1970); Leebron (1991); Post (1990); Etzioni (1999). In spite of this, the right to privacy, although redefined as a personality right, remained largely based on the individual dimension (Warren and Brandeis 1890). Neither the notion of decisional privacy nor its constitutional dimension, originating in the ground-breaking opinion given by Brandeis in his role as Supreme Court judge,[4] abandoned the individualistic nature of the right.

---

[1] See below Sect. 8.2.

[2] It should be noted that these extensive analyses are also possible without directly identifying data subjects. See also Ohm (2010); Golle (2006); Sweeney (2000a, b).

[3] In order to briefly describe the potential negative consequences of data processing at group level, it should be mentioned the potential impacts on social surveillance and the risks of group discrimination or stigmatization. See The White House (2014) and Bygrave (2002).

[4] See Brandeis' opinions in Olmstead v. United States, 277 U.S. 438, 471 (1928). See also Sweezy v. New Hampshire 354 US 234 (1957); NAACP v. Alabama 357 US 449 (1958); Massiah v. U.S. 377 US 201 (1964); Griswold v. Connecticut, 381 US 479 (1965); Roe v. Wade 410 US 113 (1973).

On the other side of the Atlantic, the notion of privacy was not influenced by the overseas experience, but was independently shaped by legal scholars and the courts.[5] Nonetheless, the protection of individual privacy was induced by the same social factors (i.e. the invasive attitude of the "penny press" and new media) that justified the response of the U.S. legal system to privacy invasion and the protection of the right to be let alone (Schudson 1978).

From the theoretical point of view, the European notion of privacy was placed in the sphere of individual rights, as in the U.S., but there is a closer connection to the general theory of personality rights (Stromhölm 1967; Giesker 1905). Moreover, in Europe, the right to privacy has not acquired the wider dimension of U.S. decisional privacy and mostly refers to informational privacy. This does not mean that the right of individual self-determination with regard to government and public bodies has not been recognised in Europe, but that it rests on the different fundamental freedoms recognised by European charters and conventions, not solely on an extensive notion of privacy.[6]

Despite these differences, the nature of the right to privacy depends primarily on the individual rights model on both sides of the Atlantic (Bygrave 2004). The collective dimension of the right has been recognised both in the U.S. and in Europe, but only as an aggregation of individual privacy issues and not as an autonomous dimension.[7]

The same considerations can be applied to the legal regime of personal data, which is regulated under data protection statutes. With respect to this, there is a partial overlap between privacy and data protection, since the protection of personal data regards both private facts referring to individuals and personal information that is publicly available. Nevertheless, the individual dimension is the object of legal protection also with regard to the computer-mediated representations of individuals.

[5] See, e.g., Trib. civ. Seine, 16 June 1858, D.P., 1858.3.62; see also Whitman (2004).

[6] See the influential decision adopted by the Federal German Constitutional Court (Bundesverfassungsgericht), 15 December 1983, Neue Juristische Wochenschrift, 1984. https://www.zensus2011.de/SharedDocs/Downloads/DE/Gesetze/Volkszaehlungsurteil_1983.pdf?__blob=publicationFile&v=9. Accessed 25 June 2014.

[7] See *inter alia* Article 29 Data Protection Working Party. 2013. Letter to Mr. Larry Page, Chief Executive Officer http://ec.europa.eu/justice/data-protection/article-29/documentation/other-document/files/2013/20130618_letter_to_google_glass_en.pdf. Accessed 27 February 2014; Irish Data Protection Commissioner. 2012. Facebook Ireland Ltd. Report of Re-Audit http://dataprotection.ie/documents/press/Facebook_Ireland_Audit_Review_Report_21_Sept_2012.pdf. Accessed 27 February 2014; Italian Data Protection Authority. 2013. Injunction and Order Issued Against Google Inc. http://www.garanteprivacy.it/web/guest/home/docweb/-/docweb-display/docweb/3133945. Accessed 27 February 2014. Only in a few hypotheses, this collective dimension is recognised as autonomous and different from the individual one. This happens in labour law, where the representatives of employees concur on the adoption of the decisions concerning surveillance in the workplace on behalf of the workers, accepting limits to privacy in these contexts. See, European Commission. Undated. Second stage consultation of social partners on the protection of workers' personal data, 7, 10, 16–17 http://ec.europa.eu/social/main.jsp?catId=708. Accessed 10 January 2015. See also specific references to the provisions of European national labour laws in Freedland (1999); Hendrickx (Undated). See also Article 4 of the Italian labour statute (L. 300/1970).

Although data protection regulations have drawn their origins from citizens' concerns about government social control, regarding the society at large (Bennett 1992; Mayer-Schönberger 1997), statutory provisions mainly concern the data subject and her/his rights. This does not mean specific provisions of the law and decisions adopted by courts or data protection authorities (hereafter DPAs) do not take into account the collective concern[8] that originated data protection statutes. Nonetheless, collective interests have been mainly protected as a sum of multiple individual issues, and collective concerns have been addressed using remedies mostly based on individual rights and their enforcement.[9]

This "architecture" of privacy and data protection, which has its roots in the model of individual rights, probably represents the main reason for the few contributions by privacy scholars on group privacy and the collective dimension of data protection. Against this background, the first studies on this argument were mostly related to the traditional notion of privacy.

In this sense, group privacy has been considered the expression of the right to privacy with regard to the information shared within a group by its members (hereafter, first approach) (Bloustein 1977, 1978) or as an autonomous manifestation of privacy referring to collective entities, which concerns their self-determination and control over information (hereafter, second approach) (Westin 1970).

According to the first approach, there is no autonomous right to privacy regarding groups, but only a peculiar attitude of individual privacy in the group context. This approach to group privacy focuses on the morphology assumed by the right to privacy when it refers to the privacy of the facts or ideas expressed by the members of a group in the group environment (e.g. privacy of association, marital privilege). Group privacy provides a guarantee that this information will not be revealed outside the group.[10]

Individual privacy describes the conditions under which a "right to be let alone" should be recognised, while group privacy determines the type of personal information sharing that goes on within a group (Bloustein 1978). Group privacy is therefore related to the private facts of the life of a group and of its members. It protects them against unlawful intrusion. In this sense, the theoretical perspective remains focused on the individual right to privacy.[11]

---

[8] See Westin 1970; Breckenridge 1970; Solove 2008; Brenton 1964; Miller 1971; Mayer-Schönberger 1997; Secretary's Advisory Committee on Automated Personal Data Systems. 1973. Records, Computers and the Rights of Citizens. http://epic.org/privacy/hew1973report/. Accessed 27 February 2014.

[9] See above fn. 7.

[10] See Bloustein (1978). In the description of the different contexts in which the right to privacy is relevant with regard to the group dimension, the author considers marital, priest-penitent, lawyer-client and physician-patient relationships. In these contexts, the right to privacy is mainly related to intimacy and secrecy.

[11] See Bloustein (1978: 125): "Group privacy is an extension of individual privacy. The interest protected by group privacy is the desire and need of people to come together, to exchange information, share feelings, make plans and act in concert to attain their objectives". This notion of group privacy focuses on secrecy and intimacy and, for this reason, it is fundamentally based on the level of trust existing among the members of a group. The consequence is a duty of confidentiality. The right concerns the nature of this duty and the breach of this obligation.

From a slightly different point of view, the second approach describes group privacy as concerning an interest of the group as such, which regards the protection of facts, acts or decisions that concern its internal affairs and its organisational autonomy. Although the holistic dimension of the group is more evident in this approach, the notion of group privacy still relies on confidentiality, which – in this case – is more closely connected to the secrecy of the activity of the group than to the secrecy of the information shared within the group by its members.[12]

Both these different interpretations of group privacy seem to be based on an individualistic idea of privacy, which concerns given subjects (i.e. the members of the group) or the group itself as autonomous collective entity. In all these analyses, the architecture of the right does not seem to be inspired by the idea of a collective, non-aggregative and super-individual interest.

It should be noted that these interpretations are consistent with the studies on group theory and have been probably influenced by them. The different approaches of legal scholars seem to reflect the more general controversy between individualistic and organic sociological theories about the nature of groups.

On the one hand, attention to the individual dimension of privacy and the interactions between different individuals (Bloustein 1978) is consistent with the notion of a group as the sum of the relationships existing among its members (individualistic theory).[13] On the other hand, when the analysis takes into consideration the information concerning the group itself as a whole (Westin 1970), the group is seen as an autonomous unit that assumes the form of an organised collective entity (organic theory).

Against this background, which represents the traditional legal framework of group privacy, recent studies have adopted a different perspective (Bygrave 2002).[14] According to this interpretation, group privacy protects information referring to collective entities – both legal persons and organisations or groups without a formal and independent identity – and acts as an extension of individual data protection to these entities. This approach – which is consistent with the organic theory – challenges the traditional idea of group privacy, which is based on the model of individual rights. Furthermore, it suggests the adoption of specific safeguards for collective interests.

In this vein, the following sections discuss the collective dimension of privacy and data protection in the context of big data analytics. Although extended collec-

---

[12] See Westin (1970). Moreover, the author points out the dimension of privacy concerning the communications among different groups.

[13] The dynamics related to group privacy draw their origin from individuals, who are aware of their level of interaction and of the related social or legal consequences (Bloustein 1978). Therefore, group privacy becomes the aggregation of individual rights in the specific context of a group. This approach is consistent with sociological individualistic theories, which consider the group as an aggregation in which individuals interact with each other in a continuous and relatively stable manner. Moreover, the members of a group have the consciousness of being part of a group and usually the group is also recognised as an autonomous social structure by third parties.

[14] On the debate regarding the application of privacy concept to collective entities, see Bygrave (2002).

tions of data and data mining are not new, the complexity and obscurity of data processing, as well as the social impact of data-driven decisions, lead policy makers and legal scholars to define new remedies to protect the collective interests of the individuals grouped by data gatherers.[15]

8.2.1 A Different Approach

As described in the previous section, the traditional approaches to group privacy show their internal coherence: the privacy of the facts or ideas expressed by the members of a group in the group environment and the protection of the information about a group are respectively consistent with the individualistic and organic theories about groups. These theories, although different, are both based on members' awareness of being part of a group and on the social dimension of the group as a network of relationships among its members.

In the light of the above, the traditional approach to group privacy considers groups that are based on stable and socially recognized relationships between individuals, although they can be informal in nature (e.g. love affairs, priest-penitent relationships) or last only for a certain time (e.g. marital relationships, association).

The present contribution shows a different point of view, which is neither focused on individual privacy nor on the idea of groups as collective autonomous entities. In the context of big data, the perspective adopted here shifts the analysis from groups to clusters of individuals, from individual rights to diffuse interests, from group privacy to collective privacy.

This view differs from the theoretical framework proposed by legal scholars in shaping the notion of group privacy,[16] but is necessary to give a specific answer to the issues arising from the present and future scenarios of the infosphere (Floridi 2013, 1999).[17]

[15] It should be noted that only few data protection laws take into account the issues related to group privacy, mainly in terms of protection of information about legal entities. See Article 4 (original wording) of the Italian Data Protection Code (D. Lgs. 196/2003) (" 'data processor' shall mean any natural or legal person, public administration, body, association or other agency that processes personal data on the controller's behalf"). The article was amended in 2011, deleting any reference to legal persons. See also Article 2(4) of the Austrian data protection law, Datenschutzgesetz 2000 – DSG 2000 (" 'Data Subjectì' ['Betroffener']: any natural or legal person or group of natural persons not identical with the controller, whose data are processed (sub-para. 8)").

[16] See above Sect. 8.2.

[17] See Floridi (2013) and Floridi (1999).

8.2.2 Big Data Analytics and Collective Privacy

Nowadays, new technologies and powerful analytics make it possible to collect and analyse huge amounts of data. In many cases, the general purposes of this new concentration of control over information[18] no longer concern single persons,[19] but adopt a large-scale perspective. Analytics investigate attitudes and behaviour of large groups, communities, and even entire countries.

Moreover, these new forms of analysis do not necessarily investigate pre-existing groups. Groups are created by data gatherers selecting specific clusters of information. Data gatherers shape the population they set out to investigate and collect information about different people who do not know the other members of the group and, in many cases, are not aware of the consequences of their belonging to a group.

These new forms of aggregation differ from the traditional idea of a group and are used by data gatekeepers to take decisions that involve the members of these clusters of people and affect their internal dynamics, with consequences for the collective issues of the people involved.

This scenario makes it necessary to consider a wider field of analysis, which is represented by the diffuse interests of the individuals that have their personal data collected, analysed and clustered in groups and categories. In this sense, protection against intrusion into private life and control over personal information still represent the main issues that should be addressed also in the big data context. Nevertheless, the different nature of the group requires a different approach, not exclusively based on individual rights.

Furthermore, the existing regulations and case laws are inadequate to address the issues arising from this change of paradigm in social investigation. For this reason, this new scale of data collection and in-depth analysis require an additional layer, represented by groups' need for the protection of their privacy and their (aggregate) personal information.

The issues relating to privacy that arise from this new situation are different from the issues of individual privacy and group privacy. We are neither in the presence of forms of analysis that involve only single individuals, nor in the presence of groups in the traditional sociological meaning of the term, given the members' lack of awareness of themselves as part of a group and the lack of interactions among people grouped into various categories by data gatherers.

Nonetheless, the collective issues related to mass profiling should be taken into consideration, given the impact they have on society and individuals. This leads us to define a new dimension of privacy (i.e. collective privacy), which has its roots in individual privacy and shares some similarities with group privacy, but differs from both these previous notions. In this sense, collective privacy does not necessarily concern facts or information referring to a specific person, as with individual privacy

[18] See also Mantelero (2014a).

[19] Big data analytics identify patterns in collective behaviours, also without identifying single individuals.

and data protection.[20] Nor does it concern aggregations of individuals that can be properly considered as groups.[21]

8.2.3 *The Collective Dimension of Privacy and the Related Interests*

In order to define the nature of the collective dimension of privacy and data protection, an aspect that should be taken into consideration concerns the importance of this collective dimension in the legal system.

The question is whether a legal notion of collective privacy is necessary and whether this collective dimension requires the granting of specific collective rights,[22] which are different from the rights already existing in the field of privacy and data protection.

First, it should be noted that, in the field considered here, we are in the presence of collective rights and not merely rights of a set of persons.[23] This is because the clusters created by the data gatherers represent a small part of a wider group of individuals with common behaviour and collective issues, which are more generally present in the society and find a mere concretisation in the specific clusters.

This is evident in commercial strategies that adopt different approaches to specific individuals, who are part of one or more categories unveiled by data analytics.[24] In these cases, given the plurality of segments into which society is divided through big data analytics,[25] seemingly innocuous classifications can assume a discrimina-

[20] In many cases private companies and governments have no interests in profiling single customs or citizens, but are interested in the attitudes of clusters of individuals. Their main goal is to predict future behaviours of given segments of population and, consequently, actively act to reach economic or political purposes. See Bollier (2010).

[21] As mentioned before, the notions of (individual) privacy and data protection have an influence on the definition of the boundaries of the collective dimension of privacy, but the larger scale affects the morphology of the related interests and their enforcement. At the same time, the notion of group privacy as hitherto described by legal scholars represents the dimension of privacy that is closer to the idea of collective privacy. For this reason, previous theoretical studies on group privacy can provide further elements to define a new set of rules to protect the collective dimension of privacy.

[22] Criticisms about the notion of collective privacy have been expressed by Vedder (1997).

[23] See Newman (2004: 128): "We can distinguish a collectivity from a set. A set is a collection of persons that we would identify as a different set were the persons included in the set to change. A collectivity is a collection of persons such that we would still identify it as the same collectivity were some or all of the persons in the collectivity to change (provided that the collectivity continued to meet some other conditions) and such that the persons who are in the collectivity identify themselves in some non-trivial way as members of this collectivity".

[24] In this sense, a commercial discrimination that affects a given set of users is relevant due to the fact that the set represents a small portion of consumers, which, in general, have a collective right not to be discriminated in negotiations.

[25] See Federal Trade Commission (2014).

tory nature in given contexts and this risk of discrimination represents a collective issue.[26]

Second, it should be pointed out that interests concerning individuals and groups are not necessarily related to the same issues at individual and collective levels. For this reason, collective rights are not necessarily a large-scale representation of individual rights and related issues.

An example in this sense is provided by credit scoring models based on big data, which predict the credit risks of individuals that live in a small geographic area.[27] These individuals are classified according to their social context in a way that bears no relationship to their individual conditions, but is based on the aggregate score of the area. If we consider this case from the perspective of individual data protection, there may be persons that have no interest to limit the use of their information. Nonetheless, at a collective level, there is a general interest of the people of a certain area to avoid to be stigmatized as potential defaulting debtors, which may represent a biased evaluation and become a source of social discrimination.

In the light of the above, collective privacy protects non-aggregative collective interests (Newman 2004),[28] which are not the mere sum of many individual interests.[29] To clarify this assumption is necessary to briefly point out that interests may be shared by an entire group without conflicts between the views of its members (aggregative interests) or with conflicts between the opinions of its members (non-aggregative interests). If the group is characterised by non-aggregative interests, the collective nature of the interest is represented by the fundamental values of a given society (e.g. environmental protection).[30]

Regarding privacy and data protection, it is difficult to imagine a common and convergent interest among the member of the groups that are analysed using big data analytics, since different people can have different opinions about the balance between the conflicting interests (e.g. extensive group profiling for commercial purposes can be alternatively passively accepted, considered with favour or perceived

---

[26] For example, the fact that a consumer belongs to a data segment for "Biker Enthusiasts" give him/her more chance to receive consumer coupons from motorcycle dealerships, but the same information may have a negative impact on his/her insurance profile, due to the supposed high probability to be engaged in risky behavior. See Federal Trade Commission (2014): "Similarly, while data brokers have a data category for "Diabetes Interest" that a manufacturer of sugar-free products could use to offer product discounts, an insurance company could use that same category to classify a consumer as higher risk".

[27] These individuals are divided into clusters on the basis of information retrieved from dozens of different sources and using hundreds of variables for their assessment. See Dixon and Gellman (2014).

[28] See Newman. Collective Interests, 131.

[29] On the contrary, an aggregative approach seems to be consistent with the notion of group privacy described by Bloustein (1978).

[30] This distinction between aggregative and non-aggregative interests is made by Newman (2004), who defines these two categories of interests respectively as "shared" and "collective" interests. As observed by Finnis (1984), a collective interest in which the contrast is attenuate may become a shared interest.

as invasive and potentially discriminatory).[31] Despite these differences, in our society there are some values that are generally considered as fundamental, such as, for instance, equality and freedom. In this sense, consumers do not accept profiling practices for discriminatory purposes and citizens are not in favour of a surveillance state that extensively reduces individual and collective privacy).

From this perspective, in a given historical and social context, there are some collective interests related to privacy and data-protection that are considered relevant in the general interests in spite of individual different opinions.[32] In this sense, collective privacy protects non-aggregative collective interests.

Regarding the interests that should be considered in respect of the collective dimension of privacy and in comparison with the traditional notion of group privacy, there is a shift of the focus from confidentiality (Bloustein 1978)[33] and control over personal information (Bygrave 2002)[34] to the issues that revolve around the risks of discrimination and the negative outcomes of massive analysis of personal data (e.g. social surveillance).[35]

Against this background, collective privacy can be described as the right to limit the potential harms to the group itself that can derive from invasive and discriminatory data processing.[36] According to this interpretation, the collective dimensions of privacy and data protection mainly regard the use of information (Cate and Mayer-Schönberger 2013; Mantelero 2014b). The source of concern is not the lack of secrecy and intimacy, which represents the object of group privacy (Bloustein 1978),[37] but the unfair and harmful use of data that is processed by using modern analytics.[38]

---

[31] The same divergence of interests exists with regard to government social surveillance for crime prevention and national security.

[32] In the light of the above, the rights related to the collective dimension of privacy assume the nature of "group-differentiated rights", which are held by members of groups on account of their group membership (Kymlicka 1995). From this perspective, groups have not only collective interests that represent the aggregation of individual interests, but also different specific interests focused on the group itself rather than on each of its members.

[33] Bloustein. *Individual and Group Privacy*, 123–186.

[34] Bygrave. *Data Protection Law*, 173–298.

[35] An example is represented by the predictive software adopted by U.S. police departments to predict and prevent crimes on the basis of extensive collection of information about previous crimes. Regarding this big data application, there have been cases in which people were enrolled in the lists of potential offenders due to merely remote connections with authors of serious crimes (Gorner 2013; Perry, W.L. et al. 2013; Koss 2015; Mantelero and Vaciago 2014).

[36] Nevertheless, this collective interest, which is relevant with regard to collective privacy, is not necessarily a shared interest. As mentioned above, single data subjects may accept invasive scrutiny of their behaviours to receive more customised services or for security reasons.

[37] Bloustein. *Individual and Group Privacy*, 182.

[38] It should be noted that big data analytics can also extract predictive inferences and correlations from publicly available information and from data voluntarily disclosed by data subjects. On the risks related to the interplay between private (commercial) surveillance and public surveillance conducted by government agencies, see also Mantelero and Vaciago (2014).

8.3 The Representation of Group Interests and Conflicting Interests

In order to define how to represent the collective interests described in the previous section, it is useful to briefly consider the fields in which the group dimension of privacy is already known, although in traditional contexts that are not characterised by extensive data collection and use of analytics.

The collective dimension of rights and the existing dualism between the individual and group are known in labour law,[39] where – under certain circumstances – trade unions and employees' representatives concur in the adoption of decisions that regard the employees and their right to privacy in the workplace.[40] The adoption of these decisions at a collective level is based on the assumption that, in some cases, the power imbalance that characterises workplaces puts the employee in a position of lack of awareness of the purposes and consequences of employer's policies (e.g. workplace surveillance). This imbalance makes it difficult for workers to take a position against illegal processing of their data.

On the other hand, the entities that represent collective interests not only have a more complete vision of the impact of specific policies and decisions, but are also less affected by situations of power imbalance. Moreover, in many cases, some policies and forms of control tend towards discriminatory actions that can affect individual workers, but have the whole group as the main and final target.

This kind of representation of collective interests is also adopted in other fields, such as consumer protection and environmental protection. These contexts are all characterized by situations of power imbalance, which affect an individual (employee, consumer, and citizen), due to disproportionate imbalance of strength between the parties (employer vs. employee, big corporation vs. consumers and citizens). Furthermore, in many cases the conflicting interests are referring to contexts where the use of IT technologies makes it more difficult to be aware of their potential negative implications.

The same situation of imbalance is present in big data contexts. Data subjects are often not aware of the fundamental aspects of data processing and are unable to negotiate their personal information. For this reason, in this field, a solution can be adopted that is similar to those described above, which are based on the role played by entities that represent collective interests.

Nevertheless, employees are part of a specific group of people, which is characterised by the relationship with the same employer; consequently, they are aware of their common identity and have mutual relationships. On the contrary, in big data contexts, the common attributes of the group often become evident only to the data gatherer (Bygrave 2002).

At the individual level, data subjects are not aware of the identity of the other members of the group, have no relationship with them and have a limited perception

[39] See the Italian Statute of the Workers' Rights, Articles 4 and 8, Act 300, 20 May 1970.

[40] See above fn. 7. See also Bygrave and Schartum (2009)

of the group's issues. Furthermore, the groups shaped by analytics have a variable geometry, since clusters of individuals can be moved from one group to another.

These aspects do not undermine the idea of a representation of collective privacy interests. On the contrary, this atomistic and fragmented dimension demands a collective representation. At the same time, the nature of the groups examined here does not make it possible to have representatives designated by the members of the group, as is the case in other contexts (e.g. workplaces).

In this sense, there are some similarities with consumer law, where there are collective interests (e.g. product security, fair commercial practices), but the parties potentially harmed by unfair practices are not connected to each other. Consequently, individual legal remedies should be combined with collective remedies[41] in a context where independent authorities responsible for consumer protection play an active role.[42]

Associations that act to protect collective interests can facilitate the response to unfair practices and be involved in a multi-stakeholder process to assess the risks related to the specific use of big data. Nonetheless, the involvement of these entities requires *ad hoc* procedural criteria in order to define which entities may act in the collective interest.[43] This designation is more difficult in the big data context where data gatherers create variable groups. In this scenario, the assessment of the social and ethical impact of the use of analytics represents, in many cases, the moment in which it is clear how data processing affects collective interests and, consequently, provides the opportunity to identify the potential stakeholders.

For these reasons and in order to actively tackle the risks related to big data, an approach based on prior risk assessment seems to be more effective to prevent dangerous hidden forms of data processing. Nonetheless, how to protect collective interests is mainly a matter of decision for policymakers. Different legal systems and different balances between the components of society can lead to different solutions.[44] This makes it difficult to identify the independent authority responsible for the protection of collective interests, which should supervise the assessment procedures.

Various countries have independent bodies responsible for the supervision of social surveillance, as well as bodies that develop anti-discriminatory policies. This responsibility is spread across different authorities that take different approaches,

[41] On the role of group actions, in order to protect individual and collective interest concerning personal information, see Bygrave (2002).

[42] It should be noted that, in the field of big data analytics, the partially hidden nature of processes and their complexity make often it difficult to bring timely class actions, unlike the case of product liability, where the nature of the damage is more evident and this facilitates the reaction of the victims. As demonstrated by the recent NSA revelations, people are not usually aware of being under surveillance, and only the leak of information can disclose this practices and open a debate on their legitimacy, as well give the chance for individuals to bring legal actions. See also European Parliament (2014).

[43] See Article 80 Regulation (EU) 2016/679.

[44] In this sense, a major favor for a marked-oriented society or for government surveillance deeply affects the quantity and quality of remedies provided by the law.

have different resources, and use different remedies. Furthermore, these authorities often do not cooperate in solving cases with multiple impacts.

Against this background, the analysis of data processing necessarily plays a central role in assessing the risks related to big data analytics. This analysis represents the element that is common to all these situations, regardless of the nature of the potential harm to collective interests. For this reason, DPAs – more than other authorities – can play a key role in risk assessment.[45]

Although DPAs do not mainly focus on specific social implications of the use of data (e.g. discrimination); rather, they focus on the main and common aspect, which is data processing. On the other hand, the adoption of an approach based on the various negative effects of the use of big data (discrimination, unfair consumer practices, social control, etc.) involves different entities and authorities. As discussed above, this may have as a final result a fragmented and potentially conflicting decisional process, with little attention to the common core, which is represented by the use of data.

At the same time, DPAs are also accustomed to address collective issues and have already demonstrated that they consider both the individual and the wider collective dimension of data processing.[46] Finally, DPAs adopt decisions that consider the procedural or security aspects of data processing, but also balance the conflicting interests that revolve around the use of data.

The adequacy of the solution proposed here is empirically demonstrated by important cases decided by DPAs concerning data processing projects with significant social and ethical impacts. These cases show that decisions to assess the impact on society of innovative products or services are not normally on the initiative of the data subjects, but primarily on that of the DPAs, who were aware of the potential risks of these innovations.[47] Moreover, these authorities are in the position to suggest measures to be adopted by companies to reduce these risks. They are also able to place this problem within the more general framework of the rights of the individual, as an individual and as a member of a democratic society.

In this scenario, assessing the impact of the use of big data plays a central role to protect collective rights.[48] Entities that represent collective interests should be able to exercise the right to participate in the processes of risk assessment,[49] which should adopt a multi-stakeholder approach.[50]

---

[45] See below Sect. 8.5.

[46] See above fn. 7.

[47] See above fn. 7.

[48] See below Sect. 8.5.

[49] In their role of representatives of collective interests, these entities could also bring legal actions for non-pecuniary damages, as well as they should be able to exercise the traditional individual rights on behalf of data subjects. See also Article 80 Regulation (EU) 2016/679.

[50] In this sense, the stakeholders may have the right to access to the documents that describe the architecture and general purposes of big data processing. Nevertheless, in order to protect the legitimate interests of companies and governments, DPAs can limit this disclosure to third parties. In the big data context, these issues are also related to the transparency of the algorithms used by companies (Citron and Pasquale 2014). See also Mayer-Schönberger and Cukier (2013), who

8.4 The Balancing Test of Conflicting Interests and the Risk Assessment

The increasing use of predictive analytics in decision-making processes, which affect groups of individuals in different fields, requires the analysis of the social and ethical consequences related to these forms of data processing and the balance of the different interests that become relevant.[51]

Since ethics and society are context-related, historical and geographical variables influence this balance, which is also the result of the strategies adopted by different policymakers. For these reasons, it is not possible to define a balance between conflicting interests that is extensively accepted in any cultural context.[52] Consequently, a balancing test should focus on a specific social context in a given historical moment[53] and, as it has been pointed out, a prescriptive general ethical guidance is problematic (Wright 2011).

Given this variability, from a theoretical perspective, the values recognised by the international charters of fundamental rights can provide a common framework for the balancing test. They may represent a baseline to identify the values that serve as an ethical guidance and to define the existing relationships between these values (Wright 2011).

The definition of the context-related values and the consequent relationship between the conflicting interests and rights should be then tailored to the specific use of big data analytics. This different "in-context" balance of the conflicting interests is based on an impact assessment that prevents negative effects of a given use of big data.

This should lead lawmakers to introduce a prior impact assessment of big data applications. This assessment should not only focus on data protection (data protec-

---

suggest a model based on independent internal and external audits. A wider access to the logic of the algorithms is required by Article 29 Data Protection Working Party. 2013. Opinion 03/2013 on purpose limitation, 47. http://ec.europa.eu/justice/data-protection/article-29/documentation/opinion-recommendation/files/2013/wp203_en.pdf. Accessed 27 February 2014. See also Gillespie (2014).

[51] See Schwartz (2011); Wright (2011); Floridi (2014); Nissenbaum (2010); Calo (2013); Dwork and Mulligan (2013); Bygrave (2002); Cohen (2013); Hofmann (2005); Richards and King (2013); Article 29 Data Protection Working Party. 2014. Statement on the role of a risk-based approach in data protection legal frameworks, 4. http://ec.europa.eu/justice/data-protection/article-29/documentation/opinion-recommendation/files/2014/wp218_en.pdf. Accessed 27 February 2014.

[52] In this sense, for example, there are legal systems that give broad relevance to national and security interests, which in many cases prevail over individual and collective privacy. On the contrary, there are countries where extensive forms of social surveillance are considered disproportionate and invasive.

[53] See, e.g., the different attitude of U.S. government with regard to surveillance, before and after the September 11, 2001 terrorist attacks. See also Bygrave (2004).

tion impact assessment)[54] and security, but it should be a multi-criteria risk-analysis,[55] which considers the ethical[56] and social consequences of data processing.[57]

In the presence of complex data collection and information processing systems affected by lock-in effects, such an impact assessment cannot be conducted either by consumers, or by companies. It requires experts in data protection law and external auditors with specific and multi-disciplinary skills.[58] For these reasons, data protection authorities can play a relevant role in this case-by-case assessment. This suggests introducing a mandatory multiple-risks assessment, which should be conducted by third parties, under the supervision of national data protection authorities, which also define the professional requirements of these third parties.[59] Furthermore, data protection authorities can involve the different stakeholders that represent the collective interests affected by specific projects of data processing in the assessment process.[60]

Obviously the entire system only works if the political and financial autonomy of DPAs, both from governments and corporations, is guaranteed.[61] Moreover, DPAs need new competences and resources to bear the burden of the supervision and approval of these multiple-risks assessments.

In the light of the above, a model based on mandatory fees — paid by companies when they submit their requests for authorization to DPAs — would be preferable.[62]

---

[54] On the data protection impact assessment see Article 35 Regulation (EU) 2016/679. It should be noted that the data protection impact assessment does not represent a new approach to data protection, as the privacy impact assessment exists in different experiences around the world and has represented an important tool since mid-1990s. On the origins of the notion of privacy impact assessment, see Clarke (2009). Nevertheless, the exiting difference between privacy and data protection necessarily affects the extent of these different assessments, which investigate fields that are not completely overlapped.

[55] On this multi-critera risk analysis, see more extensively Mantelero (2014b).

[56] See above fn. 67.

[57] See also Article 29 Data Protection Working Party. Statement on the role of a risk-based approach.

[58] It is worth pointing out that the social and ethical assessments are similar to the data protection impact assessment in their nature, since they are prior assessments based on risk analysis. Nevertheless, in these cases, the wide range of interests that should be considered requires the involvement of different stakeholders and experts.

[59] See Mantelero (2014b). The article, which deals with personal data and big data analytics, suggests adopting a new paradigm based on a mandatory multiple assessment coupled with an opt-out regime. In this model, although this assessment represents an economic burden for companies, it allows those who pass to use data for complex and multiple purposes, without requiring users to opt-in. At the same time, a prior assessment conducted by independent authorities and an opt-out model seem to offer more safeguards to users than the apparent, but inconsistent, user's self-determination based on the opt-in model.

[60] See also Wright (2011); Citron (2008). A different assessment exclusively based on the adoption of security standards or corporate self-regulation would not have the same extent and independence. This does not mean that, in this framework, forms of standardization or co-regulation cannot be adopted (Calo 2013).

[61] See also FRA – European Union Agency for Fundamental Rights (2013). See also Simitis (1987).

[62] This self-financing model, based on licensing or notification fees, was adopted in the past in Sweden and United Kingdom (Schütz 2012; Information Commissioner's Office 2011). See also the fee-based model adopted by the European Medicines Agency.

This solution provides proportionate resources to authorities without the risk that they may be influenced by the companies under their supervision. Finally, in cases of large scale and multinational data collection, forms of mutual assistance and cooperation may facilitate the role played by DPAs in addressing the problems related to the dimensions of data collections and data gatherers. Nevertheless, to achieve these goals, a subset of rules for big data analytics, focused on a multiple-risks assessment[63] and a deeper control exercised by DPAs, should be adopted.[64]

8.5 Conclusions

The analysis conducted in this chapter and the related observations represent an introductory, and mostly theoretical, overview of the issues concerning the impact of big data analytics on society. This impact leads us to reconsider the traditional concepts of group and privacy. In this sense, the previous sections suggest the adoption of the notion of 'collective privacy'.

This notion offers a different perspective that focuses on collective interests, in a context in which data processing is increasingly oriented to monitor collective behaviours, in order to predict and influence people's attitudes.

The new technological context suggests defining collective privacy as the right to limit the potential harms to the group itself that can derive from invasive and discriminatory data processing. In this light, the collective dimension regards the use of information and the use of data analytics, rather than the secrecy of information and data quality.

From this perspective, risk assessment procedures and data protection authorities play a key role in tackling the potential risks of collective harms related to unfair use of data. In respect of risk assessment, the assessment should consider both the traditional aspects regarding data protection and security, and the new issues referring to the ethical and social impact of the use of information. With regard to data protection authorities, they may have an important function in balancing the conflicting interests and in the supervision of risk assessments.

Nevertheless, to create this future framework it is necessary to define specific legal provisions for big data analytics, in order to make these guidelines effective.

[63] It should be noted that regulations that require extensive prior risk assessments under the supervision of independent authorities are not new. They are already into force in other fields that are characterized by the presence of risks for individuals and society (e.g. authorization procedure for human medicines, mandatory security standards adopted by product liability laws).

[64] For a more detailed description of the model here proposed, which also implies a review of the opt-in model in the big data context, see Mantelero (2014b).

Bibliography

Article 29 Data Protection Working Party. 2013. Letter to Mr. Larry Page, Chief Executive Officer. http://ec.europa.eu/justice/data-protection/article-29/documentation/other-document/files/2013/20130618_letter_to_google_glass_en.pdf. Accessed 27 Feb 2014.

Article 29 Data Protection Working Party. 2013. Opinion 03/2013 on purpose limitation. http://ec.europa.eu/justice/data-protection/article-29/documentation/opinion-recommendation/files/2013/wp203_en.pdf. Accessed 27 Feb 2014.

Article 29 Data Protection Working Party. 2014. Statement on the role of a risk-based approach in data protection legal frameworks. http://ec.europa.eu/justice/data-protection/article-29/documentation/opinion-recommendation/files/2014/wp218_en.pdf. Accessed 27 Feb 2014.

Bennett, C.J. 1992. *Regulating privacy: Data protection and public policy in Europe and the United States*. Ithaca: Cornell University Press.

Bloustein, E.J. 1977. Group privacy: The right to huddle. *Rutgers School of Law* 8: 219–283.

Bloustein, E.J. 1978. *Individual and group privacy*. New Brunswick: Transaction Books.

Bollier, D. 2010. *The promise and perils of big data* Washington, DC: Aspen Institute, Communications and Society Program http://www.aspeninstitute.org/sites/default/files/content/docs/pubs/The_Promise_and_Peril_of_Big_Data.pdf. Accessed 27 Feb 2014.

Breckenridge, A.C. 1970. *The right to privacy*. Lincoln: University of Nebraska Press.

Brenton, M. 1964. *The privacy invaders*. New York: Coward-McCann.

Bygrave, L.A. 2002. *Data protection law. Approaching its rationale, logic and limits*. The Hague: Kluwer Law International.

Bygrave, L. 2004. Privacy protection in a global context. A comparative overview. *Scandinavian Studies in Law* 7(319): 319–348.

Bygrave, L.A., and D.W. Schartum. 2009. Consent, proportionality and collective power. In *Reinventing data protection?* ed. Serge Gutwirth et al., 157–173. Dordrecht: Springer.

Calo, R.M. 2013. Consumer subject review boards: A thought experiment. *Standford Law Review Online* 66: 97–102.

Cate, F.H., and Mayer-Schönberger, V. 2013. *Data use and impact. Global Workshop* The Center for Information Policy Research and The Center for Applied Cybersecurity Research, Indiana University, http://cacr.iu.edu/sites/cacr.iu.edu/files/Use_Workshop_Report.pdf. Accessed 27 Feb 2014.

Chamoux, J. 1981. Data protection in Europe: The problem of the physical person and their legal person. *Journal of Media Law & Practice* 2: 70–83.

Citron, D.K. 2008. Technological due process. *Washington University Law Review* 85(6): 1249–1313.

Citron, D.K., and F. Pasquale. 2014. The scored society: Due process for automated predictions. *Washington Law Review* 89(1): 1–33.

Clarke, R. 2009. Privacy impact assessment: Its origins and development. *Computer Law & Security Review* 25(2): 123–135.

Cohen, J.E. 2013. What privacy is for. *Harvard Law Review* 126(1904): 1933.

Dixon, P., and R. Gellman. 2014. The scoring of America: How secret consumer scores threaten your privacy and your future. 43–46, http://www.worldprivacyforum.org/wp-content/uploads/2014/04/WPF_Scoring_of_America_April2014_fs.pdf. Accessed 15 Apr 2015.

Dwork, C., and D.K. Mulligan. 2013. It's not privacy and it's not fair. *Standford Law Review Online* 66: 35–40.

Etzioni, A. 1999. *The limits of privacy*. New York: Basic Books.

European Parliament. 2014. Resolution of 12 March 2014 on the US NSA surveillance programme, surveillance bodies in various Member States and their impact on EU citizens' fundamental rights and on transatlantic cooperation in Justice and Home Affairs. http://www.europarl.europa.eu/sides/getDoc.do?type=TA&language=EN&reference=P7-TA-2014-0230. Accessed 26 Feb 2015.

Federal Trade Commission. 2014. Data brokers: A call for transparency and accountability. Appendix B. https://www.ftc.gov/system/files/documents/reports/data-brokers-call-transparency-accountability-report-federal-trade-commission-may-2014/140527databrokerreport.pdf. Accessed 14 May 2015.
Finnis, J. 1984. The authority of law in the predicament of contemporary social theory. *Journal of Law Ethics & Public Policy* 1: 115–137.
Flaherty, D. 2000. Privacy impact assessments: An essential tool for data protection. *Privacy Law & Policy Reporter* 7(5): 45 http://www.austlii.edu.au/cgi-bin/sinodisp/au/journals/PrivLawPRpr/2000/45.html?stem=0&synonyms=0&query=flaherty. Accessed 11 Nov 2014.
Floridi, L. 1999. Information ethics: On the philosophical foundation of computer ethics. *Ethics and Information Technology* 1: 37–56.
Floridi, L. 2013. *The ethics of information*. New York: Oxford University Press.
Floridi, L. 2014. *The 4TH revolution. How the infosphere is reshaping human reality*. New York/Oxford: Oxford University Press.
FRA – European Union Agency for Fundamental Rights. 2013. Access to data protection remedies in EU Member States. http://fra.europa.eu/sites/default/files/fra-2014-access-data-protection--remedies_en_0.pdf. Accessed 27 Feb 2014.
Freedland, M. 1999. Data protection and employment in the European union. An analytical study of the law and practice of data protection and the employment relationship in the EU and its member. http://ec.europa.eu/social/main.jsp?catId=708. Accessed 25 Jan 2015.
Giesker, H. 1905. *Das Recht der Privaten an der eigenen Geheimsphäre. Ein Beitrag zu der Lehre von den Individualrechten*. Zürich: Müller.
Gillespie, T. 2014. The relevance of algorithms. In *Media technologies. Essays on communication, materiality, and society*, ed. T. Gillespie, P.J. Boczkowski, and K.A. Foot, 167–194. Cambridge, MA: MIT Press.
Golle, P. 2006. Revisiting the uniqueness of simple demographics in the US population. In *Proceedings of the 5th ACM workshop on privacy in electronic society*, ed. A. Juels. New York: ACM 2006.
Gorner, J. 2013. Chicago police use 'heat list' as strategy to prevent violence. Officials generate analysis to predict who will likely be involved in crime, as perpetrator or victim, and go door to door to issue warnings. *Chicago Tribune*, August 21. http://articles.chicagotribune.com/2013-08-21/news/ct-met-heat-list-20130821_1_chicago-police-commander-andrew-papachristos-heat-list. Accessed 25 Feb 2015.
Hendrickx, F. Undated. Protection of workers' personal data in the European union, 33–35, 98–101. http://ec.europa.eu/social/main.jsp?catId=708. Accessed 18 Jan 2015.
Hofmann, B. 2005. On value-judgments and ethics in health technology assessment. *Poiesis & Praxis* 3: 277–295.
Information Commissioner's Office. 2011. Budget 2011–12. Spending plans 2012–13 to 2014–15. 2http://ico.org.uk/about_us/boards_committees_and_minutes/~/media/documents/library/Corporate/Detailed_specialist_guides/ico_budget_2011-12.ashx. Accessed 27 Feb 2014.
Irish Data Protection Commissioner. 2012. Facebook Ireland Ltd. Report of Re-Audit. http://dataprotection.ie/documents/press/Facebook_Ireland_Audit_Review_Report_21_Sept_2012.pdf. Accessed 27 Feb 2014.
Italian Data Protection Authority. 2013. Injunction and Order Issued Against Google Inc. http://www.garanteprivacy.it/web/guest/home/docweb/-/docweb-display/docweb/3133945. Accessed 27 Feb 2014.
Kohler, J. 1907. *Urheberrecht an schriftwerken und verlagsrecht*. Stuttgart: F. Enke.
Koss, K.K. 2015. Leveraging predictive policing algorithms to restore fourth amendment protections in high-crime areas in a post-wardlow world. *Chicago Kent Law Review* 90: 301–334.
Kymlicka, W. 1995. *Multicultural citizenship*. New York: Oxford University Press.
Leebron, D.W. 1991. The right to privacy's place in the intellectual history of tort law. *Case Western Reserve Law Review* 41: 769–810.
Mantelero, A., and G. Vaciago. 2014. Social media and big data. In *Cyber crime & cyber terrorism. Investigators' handbook*, ed. B. Akhgar, A. Staniforth, and F.M. Bosco. Waltham: Elsevie.

Mantelero, A. 2014a. Social control, transparency, and participation in the big data world. *Journal of Internet Law* April, 23–29.
Mantelero, A. 2014b. The future of consumer data protection in the E.U. Rethinking the "notice and consent" paradigm in the new era of predictive analytics. *Computer Law & Security Review* 30: 643–660.
Mayer-Schönberger, V., and K. Cukier. 2013. *Big data. A revolution that will transform how we live, work and think.* London: John Murray.
Mayer-Schönberger, V. 1997. Generational development of data protection in Europe. In *Technology and privacy: The new landscape*, ed. P.E. Agre and M. Rotenberg. Cambridge, MA: MIT Press.
Miller, A.R. 1971. *The assault on privacy computers, data banks, dossiers*, 54–67. Ann Arbor: University of Michigan Press.
Newman, D.G. 2004. Collective interests and collective rights. *American Journal of Jurisprudence* 49(1): 127–163.
Nissenbaum, H. 2010. *Privacy in context. technology, policy, and the integrity of social life.* Stanford: Stanford University Press, 231.
Ohm, P. 2010. Broken promises of privacy: Responding to the surprising failure of anonymization. *UCLA Law Review* 57: 1701–1777.
Packard, V. 1964. *The naked society*. New York: David McKay.
Perry, W.L. et al. 2013. Predictive policing. The Role of Crime Forecasting in Law Enforcement Operations. http://www.rand.org/content/dam/rand/pubs/research_reports/RR200/RR233/RAND_RR233.pdf. Accessed 10 Mar 2015.
Post, R.C. 1990. Rereading warren and brandeis: Privacy, property and appropriation. *Case Western Reserve Law Review* 41: 647–680.
Richards, N.M., and J.H. King. 2013. Three paradoxes of big data. *Stanford Law Review* 66: 41–46.
Schudson, M. 1978. *Discovering the news. A social history of American newspaper.* New York: Basic Books.
Schütz, P. 2012. Comparing formal independence of data protection authorities in selected EU Member States. Conference Paper for the 4th Biennial ECPR Standing Group for Regulatory Governance Conference 2012. 17, fn. 73, and 18. http://regulation.upf.edu/exeter-12-papers/Paper%20265%20-%20Schuetz%202012%20-%20Comparing%20formal%20independence%20of%20data%20protection%20authorities%20in%20selected%20EU%20Member%20States.pdf. Accessed 27 Feb 2014.
Schwartz, P.M. 2011. Data protection law and the ethical use of analytics. http://www.huntonfiles.com/files/webupload/CIPL_Ethical_Undperinnings_of_Analytics_Paper.pdf. Accessed 27 Feb 2014.
Secretary's Advisory Committee on Automated Personal Data Systems. 1973. Records, computers and the rights of citizens. http://epic.org/privacy/hew1973report/. Accessed 27 Feb 2014.
Simitis, S. 1987. Reviewing privacy in an information society. *University of Pennsylvania Law Review* 135(3): 707–746.
Solove, D.J. 2008. *Understanding privacy.* Cambridge, MA/London: Harvard University Press.
Stromhölm, S. 1967. *Right of privacy and rights of personality. A comparative survey.* Stockholm: Norstedt & Soners.
Sweeney, L. 2000a. Foundations of privacy protection from a computer science perspective. In *Proceedings Joint Statistical Meeting, AAAS, Indianapolis.* http://dataprivacylab.org/projects/disclosurecontrol/paper1.pdf. Accessed 24 Jan 2015.
Sweeney, L. 2000b. Simple demographics often identify people uniquely. Pittsburgh: Carnegie Mellon University. http://dataprivacylab.org/projects/identifiability/paper1.pdf. Accessed 24 Jan 2015.
Vedder, A.H. 1997. Privatization, information technology and privacy: Reconsidering the social responsibilities of private organizations. In *Business ethics: Principles and practice*, ed. Geoff Moore, 215–226. Sunderland: Business Education Publishers.

The White House, Executive Office of the President. 2014. Big data: Seizing opportunities, preserving values Washington, DC http://www.whitehouse.gov/sites/default/files/docs/big_data_privacy_report_may_1_2014.pdf. Accessed 26 Dec 2014.

Warren, S.D., and L.D. Brandeis. 1890. The right to privacy. *Harvard Law Review* 4(5): 193–220.

Westin, A.F. 1970. *Privacy and freedom*. New York: Atheneum.

Whitman, J.Q. 2004. The two western cultures of privacy: Dignity versus liberty. *Yale Law Journal* 113: 1151–1221.

Wright, David. 2011. A framework for the ethical impact assessment of information technology. *Ethics and Information Technology* 13(3): 199–226.

Wright, D. 2012. The state of the art in privacy impact assessment. *Computer Law & Security Review* 28(1): 54–61.

Wright, D., and P. de Hert (eds.). 2012. *Privacy impact assessment*. Dordrecht: Springer.

Wright, D., M. Friedewald, and R. Gellert. 2015. Developing and testing a surveillance impact assessment methodology. *International Data Privacy Law* 5(1): 40–53.

Chapter 9
The Group, the Private, and the Individual: A New Level of Data Protection?

Ugo Pagallo

Abstract Current trends in Open and Big Data have led certain scholars to suggest the idea of expanding the notion of the "data subject" to include the protection of data groups. Nothing precludes this expansion, however, there is a question as to the type of supra-individual right groups can be given, i.e. whether data group rights should be conceived of as rights of the group qua group or, alternatively, as complementary to the protection and enforcement of individual rights. The latter has materialized with the protection of intimate associations of large civic or business membership organizations in US law (corporate rights), and also *vis-à-vis* from the legal safeguards enshrined in Articles 7 and 8 of the EU Charter of Fundamental Rights (collective/procedural rights). This contrast entails a further distinction between the fields of privacy and data protection as in connection with the different kinds of interests, or goods, the legal systems are aiming to protect. By examining certain specific problems affecting current data protection, referred to here collectively as "data fetishism," the goal is to offer a normative standpoint upon which sides can be taken in today's debate as to any new level of data protection.

Keywords Collective rights • Corporate rights • Data protection • Group rights • Harm-principle • Informational self-determination • Legal person • Privacy

9.1 Introduction

Data protection to-date has concerned individuals rather than groups. In EU law, for example, Article 2(a) of Directive 95/46/EC states that "for the purposes of this Directive… 'personal data' shall mean any information relating to an identified or identifiable natural person ('data subject')." Similarly, in the proposal for a new data protection regulation as presented by the EU Commission in January 2012, its Article 4(1) defines the "data subject" as "an identified natural person or a natural

U. Pagallo (✉)
Law School, University of Torino, Lungo Dora Siena 100 A, 10153 Torino, Italy
e-mail: ugo.pagallo@unito.it

L. Taylor et al. (eds.), *Group Privacy*, Philosophical Studies Series 126,
DOI 10.1007/978-3-319-46608-8_9

person who can be identified, directly or indirectly, by means reasonably likely to be used by the controller or by any other natural or legal person." Notwithstanding Amendment 61 to the Commission's proposal, the EU Parliament confirmed this equalization of data subjects and natural persons on 12 March 2014 with a final legislative resolution of the new data protection regulation.

In light of the current provisions on data protection, it is noteworthy that the law already protects certain types of data that individuals have in common with other data subjects. Consider Article 8 of Directive 95/46/EC, which refers to "the processing of personal data revealing racial or ethnic origin." Likewise, contemplate Article 9(1) of the proposal for a new regulation: according to the latter, we should add to the previous protection of art. 8 on "data concerning health or sex life," the processing of genetic data, i.e. "all data, of whatever type, concerning the characteristics of an individual which are inherited or acquired during early prenatal development" (art. 4(10) of the proposal). Although the focus remains on how to protect personal data, such data is shared with other data subjects and moreover, individuals can be targeted as a member of that specific (racial, ethnic, genetic, etc.) group. This risk was realized throughout the 2010–2011 Ivorian civil war (Taylor 2016). In general terms, this threat will continue to increase with the current Open and Big Data trends, particularly as "Open Data is more likely to treat types rather than tokens and hence groups rather than individuals" (Floridi 2014). So why not expand the notion of data subject and include data groups? By granting such groups their own rights, can this legal option represent the best, or even only, way to really protect certain individual rights in the field of data protection? Why not apply this latter reasoning, that which scholars have persuasively been arguing in the field of privacy rights, namely the protection of group privacy as "an extension of individual privacy" (Bloustein 2003)?

In order to hopefully offer a comprehensive view on these issues, this chapter is divided into four sections. Section 9.2 deepens the notion of group rights and how certain authors have criticized this notion on the basis of moral arguments. The aim is to stress the limits of such reasoning in the legal domain, as it is commonly admitted that a legal subject can be an "artificial person" with rights and duties of its own. The next step of the analysis regards the type of supra-individual rights that groups can have in the legal domain. Section 9.3 introduces the difference between collective rights and corporate rights, i.e. whether data group rights should be conceived of as complementary to the protection and enforcement of individual rights or, alternatively, as rights of the group qua group. This is not simply a theoretical exercise: Section 9.4 shows how the alternative has materialized with the protection of the "intimate association" of large civic or business membership organizations in the United States (corporate rights), and the legal safeguards enshrined in Articles 7 and 8 of the EU Charter of Fundamental Rights (collective/procedural rights). Such a contrast claims a further distinction between the fields of privacy and data protection. By considering the different kinds of interests, or goods, that legal systems aim to protect, Sect. 9.5 examines some specific problems affecting the idea of complementing current rights of personal data protection with group rights. Such issues can be summarized by the opinion of those aiming to protect data qua data in the field of data protection, e.g. the EU Court of Justice's *Google v. AEPD* ruling on the "right to be forgotten" from May 2014. In accordance with the old Roman maxim "do not injure others" (*alterum non laedere*), the final goal of the chapter is to stress the shortcomings of this view of a new level of data protection. On the one hand, *pace* the advocates of the protection of data qua data, attention should be drawn to the types of prejudice, risks or threats triggered by current Open and/or Big Data, trends. What this new scenario suggests is the integration of individual rights with group rights, in order to strengthen the legal safeguards against harm provoked by other agents in the system. This form of collective right appears legitimate and even urgent in the field of data protection. On the other hand, the alternative between the collective or corporate forms that such group rights present in the traditional protection of individual's privacy, requires a normative standpoint with which we can take sides in today's debate on new levels of protection for personal data. That which may make sense in privacy law does not necessarily fit the field of data protection.

9.2 The Legitimacy of Group Rights

As mentioned above, certain scholars criticize the idea of group rights on the basis of moral arguments. In this context, this debate can be summarized in connection with two main points. The first has to do with a major concern of liberal thinkers: it regards "the power that group rights may enable a group to wield over its members," much as "the potential of group rights to rival and override the rights of individuals" (Jones 2014, n. 7). The second point insists on how groups do not meet the necessary and sufficient conditions required for properly claiming they have rights, due to for example the lack of sentience, or of deliberation, which can properly attach only to the members of the group (e.g. Kymlicka 1989).

As to the latter reasoning, we should distinguish between moral and legal arguments. Reflect on the notion of "*persona ficta et rapraesentata*" as developed by Canon Law experts beginning in the thirteenth century. For example, in the *Commentary on Digestum Novum* (48, 19), Bartolus de Saxoferrato (1313–1357) reckons that an artificial person is not really a person and yet, this fiction stands in the name of the truth, because we, the jurists, establish it: "*universitas proprie non est persona; tamen hoc est fictum pro vero, sicut ponimus nos iuristae.*" This idea triumphs with legal positivism and formalism in the mid-nineteenth century. Consider the *System of Modern Roman Law* (1840–1849), in which Friedrich August von Savigny claims that only human individuals can properly have rights and duties of their own, although it is within the power of the law to grant such rights of personhood to anything, e.g. business corporations, governments, ships in maritime law, and so on.

Drawing on this longstanding tradition, it is thus commonly admitted that group rights can legally be attached to such entities as sovereign states, corporations or organizations, which survive changes in their individual memberships. The protection of collective and group rights, after all, is the bread and butter of several scholars in the fields of constitutional law, administrative law, business law, consumer law, tax law, and even criminal law. This is not to say that the rights (and duties) of such artificial persons are not problematic and also quite different according to the legal field in question. Contemplate whether artificial legal persons should be granted the same rights that individuals possess, e.g. the 2011 decision by the US Supreme Court on a corporation's freedom of speech under the protection of the First Amendment. Likewise, scholars still discuss matters of corporate epistemology as foundational to determining their legal responsibility: here, on the basis of multiple accumulated actions by both humans and computers, we need to ascertain the actual or hypothetical information content of the corporate entity so as to determine its liability.

In light of the multiple rights and duties that artificial persons have in all fields of the law, it follows that no legal principle exists precluding the aim of expanding the notion of a data subject to groups. On the contrary, certain legal systems protect the data of artificial legal persons, such as associations and corporations, as in Austria with Article 2(8) of the national data protection act, and as was in force in Italy until the amendments of Act n. 214 from 22 December 2011. Therefore, what are the types of supra-individual rights that legal systems can protect in this way?

This question brings us back to the initial arguments debated in legal and moral theories apropos group rights, namely the power that groups may have *vis-à-vis* the rights of the individuals. The next section furthers this discussion in accordance with the legal distinction between corporate and collective rights.

9.3 Between Corporate and Collective Rights

As there is no legal principle precluding the idea of expanding the notion of a data subject to groups, the next step in this analysis concerns the two different ways by which we may conceive such a supra-individual right, namely either as a corporate or collective right. In the first case, group rights are attached to an entity, such as an organization, corporation, or state, as an individual and autonomous entity having its own rights and moreover, may hold such rights as against its own members, e.g. a state against its citizens, a university against a professor, etc. In the case of collective rights, individuals share some interests or beliefs forming them into a right-holding group: in contrast to corporate rights, however, such a group is the form through which the individuals exercise their rights, rather than those of the group qua group. This difference can be further illustrated with the wording of Article 2 of the Italian constitution, according to which "the Republic recognizes and guarantees the inviolable rights of the person, both as an individual and in the social groups where human personality is expressed." Among the rights of these social groups, we find the right "to freely profess their religious belief in any form, individually or with others" (art. 19). In addition, there are some rights that, especially in Italy, appear to be in-between corporate and collective rights, such as the rights to freely establish trade unions (art. 39) or political parties (art. 49).

These examples raise the issue of whether, and to what extent, group rights are (and should be) complementary to the protection and enforcement of individual rights. A procedural complement, for instance, is supported by the aforementioned EU proposal for a new regulation on data protection. Pursuant to Article 73(2) of the proposal, "any body, organisation or association which aims to protect data subjects' rights and interests concerning the protection of their personal data... shall have the right to lodge a complaint with a supervisory authority." Furthermore, in accordance with Article 76(1), "any body, organisation or association referred to in Article 73(2) shall have the right to exercise the rights referred to in Articles 74 [i.e. the right to a judicial remedy against a supervisory authority] and 75 [i.e. the right to a judicial remedy against a controller or processor] on behalf of one or more data subjects." Hence, the overall idea of the proposal is not to replace today's personal data protection with a privacy group regime but rather, to complement the former with a new group right to lodge complaints. Since the data subject can be targeted and her privacy infringed due to her membership in a given (racial, ethnic, genetic, etc.) data group, it makes sense to grant such a group, or "any body, organisation or association which aims to protect data subjects' rights and interests," a procedural right to a judicial remedy against the data controllers, processors or supervisory authorities. This is the viewpoint as well of the EU Court of Justice ("CoJ").[1] In contrast, what about the situation where a group claims rights of its own, i.e. a subspecies of corporate rights?

Consider the case of genetic groups comprising individuals with an inherited set of instructions that biologically define their organisms. In such cases, the distinction that changes the data protection was stressed by Art. 29 of the EU Working Party in Opinion 3/2012 on developments in biometric technologies (WP 193): "in this case, it is not important to identify or verify the individual but to assign him/her automatically to a certain category." What if the target of a privacy infringement is the group, or the category, as such? Should the law grant such a group its own rights to preserve and foster its identity?

In order to avoid any misunderstandings, it should be noted that such an "identity" is not related to the protection of any alleged natural group, even in the case of genetic groups, or of groups sharing a language, religion, etc. Rather, that which is at stake here concerns the set of ontological and epistemological predicates that cluster such a group, e.g. a category defined by some predisposition towards certain types of illnesses, behaviours, etc. From a legal viewpoint, what happens if such predicates clustering a group are abused by some of its members? Would there be any substantive differences between such a case and the aforementioned examples of states holding their own rights against their citizens, or of universities against their professors?

The short answer is yes. There are different ways in which the law can preserve and foster the identity of a group qua group, depending on such parameters, as the

[1] See the ruling of the EU Court of Justice issued 8 April 2014 (C-293/12 and 594/12), and the claims by certain Austrian and Irish organizations of being victims of a violation of their rights under Articles 7 and 8 of the Charter of Fundamental Rights. We return to this below in Sect. 9.4.

interests or rights to be protected, multiple jurisdictions, and the legal field under examination. In national law, for instance, think of the multiple kinds of anti-discrimination laws aiming at protecting the rights of groups to be treated equally in political participation, employment, consumer transactions, or regardless of sex, race, language, religion, etc. In the field of international law, consider the difference between a nation's right to self-determination and the provisions of international law against genocide, such as in Article 6 of the Rome Statute of the International Criminal Court. Hence, what about the specificity of data protection and privacy laws? What are the kinds of interests or goods legal systems may aim to protect in these latter fields?

9.4 Between Opaqueness and Transparency

So far, "privacy" and "data protection," whether referring to groups or individuals, have been used as interchangeable terms in the analysis, although this is not necessarily the case. To start with, we can summarize the many ways in which the notion of privacy has been conceived of as a condition of "solitude," "exclusion" or "secrecy" (Westin 1967; Gavison 1980; Allen 1988; etc.), with Hannah Arendt's idea of "opaqueness." In the words of *Vita Activa*, "a life spent entirely in public, in the presence of others, becomes, as we would say, shallow. While it retains its visibility, it loses the quality of rising into sight from some darker ground which must remain hidden if it is not to lose its depth in a very real, non-subjective sense" (Arendt 1958: 71). This idea of opaqueness can nowadays be grasped in informational terms, that is, according to the principles and rules aiming to constrain the flow of information in the environment, so as to keep firm distinctions between individuals and society, agents and systems. The principles and rules of the legal system determine, in other words, the degrees of "ontological friction" in the informational sphere, as "the amount of work and efforts required for a certain kind of agent to obtain, filter and/or block information (also, but not only) about other agents in a given environment" (Floridi 2006). The higher the ontological friction, the lower the degree of accessibility to personal information and thus, the stronger the protection of one's privacy and opaqueness.

This idea of privacy can entail no data processing at all, e.g. cases of "unwanted fame." Even in such cases, however, the law aims to protect the flow of information that individuals deem fair to reveal, share or transfer in a given context. The protection of an individual's opaqueness through the degrees of ontological friction in the environment of course can go hand-in-hand with the protection of groups. A first legal option is given by the notion of collective rights as illustrated above in the previous section. From this latter stance, group privacy can be presented as "an extension of individual privacy. The interest protected by group privacy is the desire and need of people to come together, to exchange information, share feelings, make plans and act in concert to attain their objectives" (Bloustein 2003: 125). Significantly, the US Supreme Court has granted this protection since its judgement in *NAACP v. Alabama* [357 U.S. 449 (1958)]: "Inviolability of privacy in group association may in many circumstances be indispensable to preservation of freedom of association, particularly where a group espouses dissident beliefs" (*op. cit.*, at 462). As an extension of individual privacy, group privacy can then be considered as a collective right, rather than a corporate right, because the aim of this protection is not to replace the legal safeguards of individual privacy with a privacy group regime but rather, to complement the former with the latter.

In a more recent case, however, i.e. *Boy Scouts of America v. Dale* [530 U.S. 640 (2000)], the US Supreme Court abandoned this view and consequently, the previous opinion that large civic or business membership organizations have no constitutional right of "intimate association" [*Roberts v. United States Jaycees*, 468 U.S. 609 (1984)]. The Court declared in *Boy Scouts* that a statute promoting associational interests of homosexuals unconstitutionally restricts privacy and associational rights, since the Boy Scouts, a large civic membership organization after all, could legitimately claim a right to associational privacy even against the excluded leader of the group. Moreover, that which the Scouts deemed as private, could intrude on the private life of the excluded group leader, making his sex life subject to retribution. Because the Scouts had clearly stated their moral preferences in their membership rules, it follows that the intimate association of the Scouts and "its" privacy, i.e. the associational privacy of the Boy Scouts as a corporate right, should prevail over "their" privacy, namely the right of the excluded group leader claiming an individual right. The Court conceded that the privacy of the group, as a single and unitary holder, can be conceived analogously with an individual's privacy.

Things are different in Europe. In *Church of Scientology of Paris v. France* (appl. 19509/92), for example, the European Commission of Human Rights admitted that "any non-governmental organisation claiming to be the victim of a violation by one of the High Contracting Parties of the rights set forth in the Convention" – i.e. the 1950 European Convention of Human Rights, or ECHR – is entitled to a judicial remedy pursuant to Article 34 (§ 2 on "the law" of the decision from 9 January 1995). However, the Commission rejected the argument of the applicant association that considered itself "entitled to a 'private sphere' and that it can invoke for itself, as a legal entity, and/or on behalf of its members, the right to respect for private life" (§ 1 of the complaints). This reasoning hinges on the difference between the right protected by Article 8 ECHR, i.e. the right to respect for private and family life, and the right safeguarded by Article 9, namely freedom of religion. In this latter case, we may admit a sort of corporate right, while in the case of privacy, we are dealing with an individual right. In the words of the Commission, "it is true that under Article 9 of the Convention a church is capable of possessing and exercising the right to freedom of religion in its own capacity as a representative of its members and the entire functioning of churches depends on respect for this right… However, unlike Article 9, Article 8 of the Convention has more an individual than a collective character, the essential object of Article 8 of the Convention being to protect the individual against arbitrary action by the public authorities" (§ 2 on "the law" of the decision).

Scholars have largely discussed whether the European Court for Human Rights ("ECtHR") should overturn this standpoint, and admit the right to complaint by groups and other "artificial legal persons" pursuant to Article 8 ECHR. This is the procedural approach of the EU CoJ and its reading of Articles 7 and 8 of the EU Charter of Fundamental Rights in C-293/12 and 594/12 on the Data Retention Directive 24/2006.[2] However, that which some scholars claim goes even further (van der Bart 2014), namely a corporate rather than collective right under the umbrella of Article 8. According to this stance, in such cases as *Big Brother Watch and others v. UK* (application 58170/13 before ECtHR), the applicants, i.e. an association, could legitimately claim to be victim of a violation of their rights under Article 8, insofar as the UK intelligence services would have been collecting and processing data in such a way that is neither "proportionate" nor "necessary" in a democratic society.

Whether or not *Big Brother Watch* wins its case against the UK government, a further difference between the rights protected by Articles 8 (collective) and 9 (corporate) of the Convention should nonetheless be stressed. It is difficult to imagine the ECtHR overturning two pillars of its case law on the kind of protection set up by the ECHR legal framework. First, in order to legitimately claim a violation of their rights, applicants, including associations, have to demonstrate that some sort of damage is involved in the case. Second, such damage never entails the protection of organizations against the members of the group but rather, the protection of the group against "its" state. That which is at stake here does not concern the protection of corporate rights for large civic membership associations, as in the US. Rather, in the field of the ECHR protection in such cases as *Big Brother Watch v. UK*, the issue has to do with a procedural right to a judicial remedy against governments and states in the sphere of private life, i.e. on the basis of personal damage suffered by some individuals, such as the applicants and members of the group. The larger such civic or non-governmental groups are, the likelier that some of their members have suffered personal injury.

Things are different in the field of data protection. Contrary to privacy's "opaqueness," issues of data protection mostly revolve around the transparency with which such data are collected, processed and used. Under EU law, individuals have the right to know the purposes for which their data are processed, as well as the right to access that data and to have it rectified. In the wording of Article 8(2) of the EU Charter of Fundamental Rights, "such data must be processed fairly… and on the basis of the consent of the person concerned or some other legitimate basis laid down by law." This type of protection through the principles of minimization and quality of data, its controllability and confidentiality, regards the one-to-one interaction between parties to a contract, or other types of agreement, such as an internet service provider (ISP)'s terms of service. Duties and rights between the data subject and the data controller of course may overlap with the protection of the individual "opaqueness." In such cases, the aim is to constrain the flow of information, and keep firm distinctions between individuals and society, in order to protect what the

[2] See above note 1 in Sect. 9.3.

German Constitutional Court has framed in terms of "informational self-determination" since its *Volkszählungs-Urteil* ("census decision") from 15 December 1983. This general right to the *informationelle selbst-bestimmung* of individuals includes the right to determine whether personal data can be collected and, eventually, transmitted to others; the right to determine how that data may be used and processed; the right to access that data and, where necessary, to keep it up to date; down to the right to delete that data and refuse at any time to have the data processed. Contrary to the US approach to corporate privacy and the views of scholars aiming to import this model in Europe, the corollaries of the BVerG doctrine on the *informationelle selbst-bestimmung* of the individual suggests a cautionary tale as to whether legal systems should expand the number and rights of data subjects to groups qua groups.

A first reason for such caution hinges on the very difference between privacy's "opaqueness" and data protection's "transparency." As shown by Articles 73(2) and 76(1) of the proposal for a new data protection regulation in Europe, mentioned above, we may welcome a new set of group rights in the field of data protection. However, these rights have to be conceived as the form through which individuals exercise their own rights, that is, as collective rights, rather than those of the group qua group. Otherwise, we allow the rights of groups to determine the conditions for legitimizing data processing, even against the will of its own members, and deciding whether personal data of such members can be collected, or transmitted to others, as well as whether such data should be deleted. Reflect on the number of cases in which the processing of personal data is legitimate regardless of the individual consent, e.g. current Article 7(d) of EU Directive 46/95/EC on "processing that is necessary in order to protect the vital interests of the data subject." By envisioning a new generation of group rights as corporate rights in the field of data protection, such cases rendering the consent of the individuals unnecessary would be multiplied without reasonable grounds. Although no legal principle precludes the idea of expanding the notion of data subject to groups, two tenets of the rule of law, such as individual autonomy and anti-paternalism, would be imperilled (Pagallo 2012) by capturing the rights of such groups as corporate rights in the field of data protection.

A further reason why we should beware of expanding the rights of individual data subjects to groups concerns some current trends in data protection. Rather than the aforementioned one-to-one interaction between parties to a contract, or other types of agreement, these trends regard the many-to-many rights and obligations of individuals' extra-contractual interaction and the role of some ISP intermediaries, such as the role of search engines (Pagallo 2011). Especially in Europe, the interest, or good, protected by the law can be difficult to detect in some cases, as can occur with a new set of duties and obligations imposed on "third parties" as a result of what the EU Court of Justice declared in *Google v. AEPD* from 13 May 2014, namely the famous case on the "right to be forgotten." Here, the Court had to determine whether the data subject has a right that the information relating to her personally cannot be linked to her name by a list of results displayed by a search engine, such as Google's, following the query made on the basis of her name (C-131/12). In the wording of the Court, “it is not necessary in order to find such a right that the inclusion of the information in question in the list of results causes prejudice to the data subject” (§ 96 of the decision). Consequently, we may conceive cases in which the protection of the rights of the data subject aims to safeguard the data as such, regardless of any harm, prejudice or privacy issue that can trigger the obligation of third parties.[3] Contrary to the privacy case law of both the ECtHR and the US Supreme Court, this no-harm doctrine of the EU CoJ, namely, the protection of data qua data, is what I term data fetishism. The following section dwells on the short-comings of this approach, which reverberate on the legitimacy of a new level of data protection.

9.5 On Data Fetishism

The idea of complementing the current rights of the personal data protection framework with new group rights suggests that attention should be drawn to the kind of interest, or good, that today’s legal systems ought to defend. Leaving aside the conceptual debate on the theory of rights, this chapter has claimed so far, based on the notions of interest, choice, legal goods and the like, that such an integration of individual rights with group rights can be seem as legitimate and even urgent today, in order to strengthen the protection of individuals against harm, prejudice, risks or threats as raised by current Open Data and Big Data trends. Rather than a unique data subject whose informational self-determination is specifically under attack, individuals will more often be targeted as a member of a group, or as a specimen falling within the set of ontological and epistemological predicates that cluster a group. New types of threats and harms should be expected as a result: in the phrasing of Floridi, it is more about the new protection of “sardines,” i.e. individuals as members of a group, than “Moby Dicks.” And while “the individual sardine may believe that the encircling net is trying to catch it… it is not… it is trying to catch the whole shoal” (Floridi 2014: 3). Correspondingly, the traditional type of protection against individual harm in the field of data protection should be supplemented with an analysis of the risks and threats to the processing and use of group data that may provoke new kinds of harm to most of us, namely the “sardines.” What is the interest, choice or good of the sardines in peril?

[3] It may be argued that once search engines are deemed data controllers, as the CoJ did in C-131/12, §§ 33–34, Google and any other search engine for that matter, should not be considered as a third party. Rather, ISP obligations would be those typically associated with the one-to-one legal interactions mentioned above in Sect. 9.4. Yet, in previous rulings, e.g. *Google vs Louis Vuitton* from 23 March 2010, it is noteworthy that the opinion of the Court was different, in that ISP obligations depended on “the actual terms in which the service is supplied” and “whether the role played by that service provider is neutral, in the sense that its conduct is merely technical, automatic and passive” (§ 114 of the decision). Admittedly, it would be interesting to examine why the CoJ Justices changed their mind. For the sake of concision when dealing with group rights as a new level of data protection, we will skip this level of analysis here.

Remarkably, this is the point that advocates of data fetishism simply overlook. In addition to the remarks of the EU CoJ in *Google v. AEPD*, contemplate the field of the reuse of public sector information (PSI) in Europe, and how some member states and public sector bodies alike often refer to current data protection safeguards as a preposterous way to curb manifold legitimate applications of PSI reuses (Pagallo and Bassi 2013). By sticking a formalistic and at times, pedantic interpretation of the legal texts, such an aim to protect data qua data, regardless of any harm or prejudice to the data subject, ends up with two major problems and a paradox.

First, privacy and data protection are not "absolute rights," but "relative rights" (Pagallo 2013). Contrary to the protection from retrospective criminal penalties or the prohibition of torture, privacy and data protection frequently entail matters of balancing. Pursuant to Article 8 of ECHR, for example, the right to privacy can be limited "in the interests of national security, public safety or the economic well-being of the country, for the prevention of disorder or crime, for the protection of health or morals, or for the protection of the rights and freedoms of others." In the ECtHR case law, this "relative" nature of the right to privacy suggests why legal protection often revolves around that "necessary" in a democratic society, or in accordance with the principles of proportionality and predictability, so as to defend individuals against arbitrary interference by public authorities. Such a balancing is similarly at work in other fields of the law, suffice it to mention, the cost-benefit approach of the economic analysis of the law to matters of tortious liability, that is, extra-contractual obligations generally imposed against the will of the party seen as invoking in some sense the harm. From this latter point of view, the focus is on the "disparity between the cost (great) of the act to the victim and the (small and even negative) cost to the injurer of avoiding the act" (Posner 1988: 868). Although we do not have to accept all of Posner's ideas, it is even more difficult to buy current opinions on how to protect the data of individuals for their own sake, or against no harm, in the many-to-many contexts of extra-contractual interactions.

Second, we may admit that any costs to the victims, harm done to the data subjects, etc., can simply be presumed. Data protection infringements, in other words, could be likened to speed limits that should be respected in spite of whether someone is driving or walking on the street in the middle of the night. This precautionary approach makes good sense in the current Big or Open Data era. We may presume that opaque, or hidden, processing of group data ends up with physical threat, injury, and lack of confidentiality, much as occurred in the 2010–2011 Ivorian civil war. However, the parallel falls short in capturing that which advocates of data fetishism really claim: the rights which they aim to protect do not hinge upon a risk-analysis that has to determine the probability of events, their consequences and costs, so as to specify, or quantify, the threat of a given behaviour and hence, the presumption of harm. Rather, that which is at stake concerns the protection of data qua data, regardless of any alleged prejudice to an individual's informational privacy or personal data. Going back to the ruling of the EU Court of Justice in *Google v. AEPD* (C-131/12), "it follows from the foregoing considerations that… Article 12(b) and subparagraph (a) of the first paragraph of Article 14 of Directive 95/46 are to be interpreted as meaning that, when appraising the conditions for the application of those provisions, it should inter alia be examined whether the data subject has a right… without it being necessary in order to find such a right that the inclusion of the information in question in that list causes prejudice to the data subject" (§ 99 of the decision).

Third, we have the paradox. By asserting that, *pace* the harm principle (e.g. Stuart Mill 1859), data protection may impose duties on third parties that require no prejudice to the data subject, advocates of data fetishism make it difficult to understand the new challenges and threats triggered by Open or Big Data trends. Think of such techniques as data mining or profiling, and their models for assembling groups in accordance with certain educational, occupational or professional capabilities, social practices (e.g. a religion), or social characteristics (e.g. an ethnicity), along with the prediction of their behaviour, in order to include or exclude the "new sardines" from a particular service, product or credit, etc. Not only can individuals be targeted as members of a group, but they can even ignore being a part of that group on the basis of a set of ontological and epistemological predicates that cluster people into multiple categories (Pasquale 2014). Accordingly, the focus should be on that which the advocates of data fetishism overlook. By taking into account the threats and risks of the new scenario brought on by current trends of Open Data, Big Data, etc., we should reflect on whether the traditional kind of protection against individual harm in the field of data protection can still disregard the analysis of the menace that the processing and use of group data raise in terms of physical threat or injury, unlawful discrimination, loss of confidentiality, identity theft, or financial loss, etc.

Notably, this is the stance partially put forward by the aforementioned proposal for a new regulation on data protection, in which the EU Commission insists on the relevance of data protection impact assessments that should determine risks and threats for the processing and use of certain kinds of personal data. Consider Article 33 of the proposal, according to which data controllers have the responsibility of performing a data protection impact assessment "where processing operations present specific risks to the rights and freedoms of data subjects by virtue of their nature, their scope or their purposes." For example, specific risks concern "a systematic and extensive evaluation of personal aspects" of the data subject (art. 33(2)(a)); "information on sex life" (art. 33(2)(b)); "monitoring publicly accessible areas" (art. 33(2)(c)); and so on. Although the EU Parliament has proposed a new article on data protection impact assessments with its Amendment 88, so that Article 25a would also refer to specific risks due to "large scale filing systems" and "measures that produce legal effects concerning the individual or significantly affect the individual," the output of this institutional work in progress is nonetheless clear. The more the focus is on how individuals can be damaged by data mining techniques, behavioural analyses, profiling, etc., the less we run the risk of guarding data qua data, the more attention is drawn to the type of interest or good the legal systems aim to protect.

Admittedly, the overall idea is nothing new as it can be traced back to the ancient Roman maxim, *alterum non laedere*, that is, "do not injure others." In criminal law, the legal accountability for this kind of behaviour has been typically imposed on individuals who voluntarily commit a wrong prohibited by law; in contracts, the idea traditionally regards compensation to those affected by the harmful behaviour of a counterparty; in tort law, payment follows from obligations between private persons usually imposed by the state to compensate for damage provoked by wrongdoing. In light of this framework, that which is new with the old Roman maxim, i.e. do not injure others, in the realm of data protection, concerns both the ways and levels at which individuals may harm one another. New techniques have put group rights in the spotlight. And although such rights can be (and often are) understood in different and even opposite ways, e.g. a procedural vs. a substantive collective right, we should not miss a crucial point in the analysis. Whether new group rights have to be added to the traditional protection of individuals is an open question hinging on the type of harm, threat or risk that individuals may cause due to their personal fault or negligence. This open question on collective vs. corporate rights of the group, however, suggests why the current debate on group rights offers a fruitful standpoint to deter any further data fetishism.

9.6 Conclusions

We have seen in this chapter that neither the legal existence of group rights nor their collective forms are issues in the fields of privacy and data protection. Think again of the Supreme Court's ruling in *NAACP v. Alabama* (1958), the decision of the European Commission of Human Rights in *Church of Scientology of Paris v. France* (1995), and Articles 73(2) and 76(1) of the proposal for a new data protection regulation in Europe (2012–2014). Against this backdrop, that which is still controversial in today's debate concerns two key points: (i) whether further collective rights should be implemented in the new EU data protection regulation and (ii) whether the US constitutional model of "intimate" corporate rights can be conveniently exported to Europe.

The first issue, i.e. additional collective rights in data protection, is empirical. Whether the collective rights of the EU proposal for a new data protection framework will be effective, or good enough, to tackle the challenges of current Open and Big Data trends, hinges on the type of harm, threat, or risk that impact assessments, such as those of Article 33 of the EU proposal, should evaluate in terms of probability of events, their consequences and costs. Here, *pace* the advocates of data fetishism, the focus should be on the prejudice that other individuals, corporations, or states may cause due to their personal fault or negligence when collecting and processing data.

The second issue, i.e. a new generation of corporate rights in data protection, is trickier. We have to distinguish between the aim to protect the data of groups not concerning their members, and the protection of data that a group may claim qua group against such members. The first scenario has been illustrated by Article 2(8) of the Austrian data protection act, and that which was in force in Italy until the amendments of Act n. 214 from 22 December 2011. The ruling of the US Supreme

Court in *Boy Scouts of America v. Dale* (2000) elucidates the second scenario: what a large civic membership organization considers as private, can impinge on the life of its members and make their intimate life subject to retribution.

The first scenario is not particularly problematic: it depends on policy considerations on the transparency with which such artificial persons, as organizations or corporations, should process their data (mind, not that of their members).

The second scenario is not problematic for an opposite reason. By admitting a new generation of corporate rights in the field of data protection, we would multiply without reasonable grounds the cases in which the consent of individuals is unnecessary, e.g. Article 7(d) of EU Directive 46/95/EC. Two pillars of the rule of law, individual autonomy and anti-paternalism, would be in peril, as the group, rather than its members, would determine whether personal data of such members can be collected, transmitted to others, or deleted, etc.

Admittedly, this mechanism of "notice and consent" as laid down by Article 7 with its exceptions, is currently under pressure: privacy notices are more often labyrinthine and it is difficult for individuals to determine the long-term risks of their consent so as to balance them against short-term gains. Additionally, as stressed by the EU data protection authorities in "The Future of Privacy" from 2009, consent appears an inappropriate ground for processing, "especially when there is a clear unbalance between the data subject and the data controller" (WP 168: 17). What reasonable solution is then at hand?

Once the corporate rights option in the field of data protection is discarded, attention should be drawn back to the types of harm, threat, or risks, raised by Open Data and/or Big Data trends. This is, after all, the standpoint of the EU regulation with a new generation of data protection impact assessments. This perspective not only deters any kind of data fetishism, but goes hand-in-hand with the collective rights laid down by Articles 73(2) and 76(1) of the proposal. Whether such rights should be advanced is a matter of empirical evidence; yet, from a conceptual viewpoint, how we should approach this new level of data protection is fairly clear. By focusing on the notion of harm and the Roman maxim "do not injure others," such others, as a group with collective rights, can at times represent the best or even the only way to truly protect the individual's right to data protection.

References

Allen, A. 1988. *Uneasy access: Privacy for women in a free society*. Totowa: Rowman and Littlefield.

Arendt, H. 1958. *The human condition*. Chicago: University of Chicago Press.

Bloustein, E.J. 2003. *Individual & group privacy*, 2nd ed. New Brunswick: Transaction Publishers.

Floridi, L. 2006. Four challenges for a theory of informational privacy. *Ethics and Information Technology* 8(3): 109–119.

Floridi, L. 2014. Open data, data protection, and group privacy. *Philosophy and Technology* 27: 1–3.

Gavison, R. 1980. Privacy and the limits of the law. *Yale Law Journal* 89: 421–471.

Jones, P. (2014, Spring). Group rights. *The Stanford Encyclopaedia of Philosophy,* E.N. Zalta, ed. http://plato.stanford.edu/archives/spr2014/entries/rights-group.

Kymlicka, W. 1989. *Liberalism, Community and Culture*. Oxford: Clarendon.

Pagallo, U. 2011. ISPs & Rowdy web sites before the law: Should we change today's safe harbour clauses? *Philosophy and Technology* 24(4): 419–436.

Pagallo, U. 2012. Cracking down on autonomy: Three challenges to design in IT law. *Ethics and Information Technology* 14(4): 319–328.

Pagallo, U. 2013. Online security and the protection of civil rights: A legal overview. *Philosophy & Technology* 26(4): 381–395.

Pagallo, U., and E. Bassi. 2013. Open data protection: Challenges, perspectives, and tools for the reuse of PSI. In *Digital enlightenment yearbook 2013*, ed. M. Hildebrand, K. O'Hara, and M. Waidner, 179–189. Amsterdam: Ios Press.

Pasquale, F. 2014. *The Black Box Society: The secret algorithms that control money and information*. Cambridge, MA: Harvard University Press.

Posner, R. 1988. The jurisprudence of skepticism. *Michigan Law Review* 86(5): 827–891.

Stuart Mill, J. (1859). On liberty. In *Collected works of John Stuart Mill*, ed. Robson, J. M. vol. 21, 259–340. Toronto: Toronto University Press (1963 ff.).

Taylor, L. (2016). No place to hide? The ethics and analytics of tracking mobility using mobile phone data. *Environment and Planning D: Society and Space* 34(2): 319–336.

van der Sloot, B. (2014). Privacy in the post-NSA Era: Time for a fundamental revision?, *JIPITEC* 5(2): para 1;

Westin, A.F. 1967. *Privacy and freedom*. New York: Atheneum.

Chapter 10
Genetic Classes and Genetic Categories: Protecting Genetic Groups Through Data Protection Law

Dara Hallinan and Paul de Hert

Abstract Each person shares genetic code with others. Thus, one individual's genome can reveal information about other individuals. When multiple individuals share aspects of genetic architecture, they form a 'genetic group'. From a social and legal perspective, two types of genetic group exist: Those which map to social groups – 'genetic classes' – and those which are perceived through interrogation of shared genetic code – 'genetic categories'. Both of these groups may be seen to have legitimate interests affected when data about them are processed. This contribution considers if these interests can be effectively protected by the Data Protection Regulation. The contribution finds that the Regulation explicitly excludes genetic groups only in a relation to a limited number of provisions. Yet, the contribution also finds that the use of the Regulation to protect genetic groups would raise significant technical and substantial problems. In light of these problems, the contribution suggests a way forward based around guidance and ex ante oversight.

Keywords Genetics • Genomics • Senetic privacy • Genetic groups • Genetic classes • Genetic categories • Genetic privacy • Group privacy • Privacy • Data protection law • Data protection directive • Data protection reform • Data protection regulation

---

D. Hallinan (✉)
FIZ Karlsruhe – Leibniz-Institut für Informationsinfrastruktur,
Hermann-von-Helmholtz-Platz 1, 76344 Eggenstein-Leopoldshafen, Germany
e-mail: Dara.Hallinan@fiz-Karlsruhe.de

P. de Hert
Vrije Universiteit Brussel, Brussels, Belgium

L. Taylor et al. (eds.), *Group Privacy*, Philosophical Studies Series 126,
DOI 10.1007/978-3-319-46608-8_10

10.1 Introduction

Increasing amounts of information can be extracted from human biological samples. Samples can be subjected to a sequencing process to produce genomic data – raw genetic data. In turn, this genomic data can be analysed to extract meaning about an individual's current, or future, physical, psychological and even social traits.

However, whilst each individual's genome is unique, this uniqueness is relative and each person may expect to share much of their genetic code with others. This means that when a genetic sample is extracted from one individual, it can also be used to produce information about others. When a group of individuals share genetic architecture, these individuals collectively might be thought of as a 'genetic group'.

It has been established that the extraction and use of genetic information may pose a significant risk to the privacy rights of the individual from whom the biological sample was extracted. The European Court of Human Rights, for example, in the *Marper* case, observed the 'intrinsically private character [of DNA]' (European Court of Human Rights 2008, §104). However there are also arguments that a separate set of privacy interests may be at risk: those relating to genetic groups.

Certain genetic groups may map to recognised social groups – for clarity, these will be referred to as 'genetic classes' throughout the article. Such groups might be argued to possess privacy interests by virtue of themselves being autonomous and self-determined entities. Other genetic groups, however, can only be perceived through the interrogation of shared genetic code and do not map to socially recognised groups – these will be referred to as 'genetic categories'. Nevertheless, each member of such a genetic category can be affected by processing related to the category. Therefore a collective interest can be seen to exist in how the category as a whole is treated.

The idea of genetic groups having privacy interests might come across as strange. However, we might recall the highly controversial history of sorting and judgments of groups of individuals based on biology, or inheritance. Learning from the lessons of the past, is it surprising that norms of behaviour toward genetic groups might exist? If not, then why should these norms not receive legal recognition?

Data protection law is the area of law which elaborates how legitimate interests in data processing are to be protected. This area of law has recently undergone reform. The currently applicable current overarching instrument – Directive 95/46 – will be replaced by the General Data Protection Regulation when it comes into force in 2018. If the argument that genetic groups also have interests engaged by data processing are taken seriously, then it would make sense to look to data protection law as an area of law through which these interests could be taken into account.

However, modern data protection law has developed with the protection of the individual in mind. The idea of protecting genetic groups is novel and their inclusion as a subject of data protection has been little considered. Nevertheless, data protection law has been designed to be flexible and to adapt to novel phenomena in data processing.

This contribution considers whether the protection outlined by the Data Protection Regulation might be extended to include genetic groups. It then considers what the consequences and problems of such an extension might be. Finally, taking both the opportunities and problems identified into account, it proposes a way forward.

The contribution begins with a brief explanation of genetics, the way in which genetic architecture might be shared and the different types of genetic group which might be recognised (Sects. 10.2 and 10.3).

It continues with a consideration of the logic behind recognising genetic groups as legitimate interest holders and as objects of legal protection (Sect. 10.4). The justification for genetic classes and genetic categories are considered separately.

Next, data protection law – and in particular the Data Protection Regulation – is explained. The regulatory approach followed by the Regulation is briefly outlined, as are the four mechanisms it takes to protecting interests in data – advance checking; organisational, procedural and technical obligations; data subject rights; oversight, compliant and redress (Sect. 10.5).

The possibility to include genetic groups as a subject of protection within each of these mechanisms is then considered (Sects. 10.6, 10.7, 10.8 and 10.9). Interestingly, a clear legislative obstruction to including genetic groups as subjects of protection only appears in relation to 'data subject' rights. Even this obstruction could be removed with a relatively small (and not unprecedented) alteration to one of the concepts used in the Regulation – the concept of the 'data subject'.

However, the contribution goes on to observe that including genetic groups as subjects of protection would not be unproblematic (Sect. 10.10). First, there are legal technical issues which would need to be resolved before the framework laid out by the Regulation would be functional in relation to genetic groups. Second, without careful consideration as to how to include genetic groups as rights holders, this would likely come at an unfair cost to other parties whose interests in the processing of data have hitherto been recognised – data subjects and data controllers.

Finally, considering both the opportunities, and the problems, discussed in previous sections, we consider a possible approach moving forward (Sect. 10.11). We advocate a soft approach. On the basis of guidance from the European level interpretation body, we suggest that the initial focus of protection should be on the *ex ante* checking mechanisms. Processing affecting genetic groups would thus be made transparent. On the basis of the information generated, Data Protection Authorities could then decide on applicable, relevant and proportionate protection for groups on a case by case basis. From here, jurisprudence related to genetic groups may develop from which more general rules may emerge.

10.2 A Brief Introduction to Genetics, Shared Genetics and Genetic Groups

Living organisms are understood to pass instructions for their own replication from generation to generation (Hartl and Ruvolo 2012). The genetic code is one of the most important biological media for the transfer of such instructions. This is contained within the DNA molecule in each cell of an organism (Aubret et al. 2004; Beisson 2008). The code consists of a chain consisting of four types of chemical – in the case of the human genome this chain is 3.2 billion letters long. The order of the chemicals and their position on the chain defines what affect they will have on the function of the cell – and therefore on the organism as a whole (Nomper 2005).

Genetic data is information which relates to this genetic code. Generally speaking, there are two types of genetic data, each of which can be read from the other. First, there is information about what a specific genetic code is (the genetic code for eye colour – genotype). Second, there is information as to what a specific architecture signifies (the resulting eye colour – phenotype). Each of these types of information can be known on an abstract level – for example, it is known that a mutation on the HEXA gene plays a role in Tay-Sachs disease.[1] They can also be applied to produce information on an individual level – for example, if it is known that John's genome displays a mutation in the HEXA gene, it is known that John possesses the genetic architecture corresponding to Tay-Sachs disease.[2]

Genetically influenced traits can vary broadly in form. It has been demonstrated that genetic code can be influential in defining certain physical characteristics (Rees 2003). It has also been proposed that it can define psychological, behavioural or social characteristics – although the strength of this connection is highly disputed (Rouvroy 2008). Genetically influenced traits may already have manifested at the moment of analysis of the genome. However, the genome may also give indications as to characteristics which may emerge in future (Article 29 Data Protection Working Party 2004).

Each individual's genome is unique. First, genes are subject to random mutations. Second, each individual will receive a history of such specificity from both father and mother – each parent will also have had a unique genome. In turn, each individual's genome is completely inalienable. It cannot be transferred or faked. Accordingly, whenever an individual's genetic data is interrogated to produce information, that information, by nature, is revelatory about that individual.

However, human reproduction is also a process of copying (it would not be called *re*production otherwise). Despite small variations, genetic information is passed down from one generation to the next with remarkable stability. Accordingly, from

[1] http://www.geneticseducation.nhs.uk/genetic-conditions-54/710-tay-sachs-disease-new. (Last consulted 27.05.2015).

[2] Further observations may be made from this genetic information. Tay-Sachs is particularly prevalent in the Ashkenazi Jewish population. If I know that John has the architecture related to Tay-Sachs, I might also assume that there is an above average chance that he is of Ashkenazi Jewish descent.

one individual's genetics, knowledge about how specific genes are passed down from parent to offspring can be used to extrapolate information about family members. The closer the blood relation, the more shared genetic architecture might be expected (Nuffield Council of Bioethics 2007).

Such extrapolation also works beyond the 'genetic family' as normally understood in social terms. The genetic family is nested within a number of broader sets of group with shared inheritance. For example, a family may also belong to a specific ethnic group, or come from a certain geographical area. Members of the family will thus share genetic architecture with other individuals with the same heritage. The genetic groups an individual belongs to can be observed at an ever expanding size until, eventually, the whole human race is encompassed (Lowe 2001).

Equally, aspects of genetic architecture might not need to be linked with inheritance to be held in common. For example, a specific single mutation may be visible across a number of people who would not normally be regarded to share a common heritage. For example, two sufferers of Huntington's Chorea may have completely different genetic backgrounds, yet would share the genetic architecture relevant to Huntington's disease sufferers.

When genetic architecture is shared across multiple individuals, this shared architecture is a commonality which can be used to classify these individuals as a 'genetic group'.

10.3 Two Types of Genetic Group: 'Genetic Classes' and 'Genetic Categories'

There are any number of groups which can be perceived through shared genetic architecture. However, when perceived through a social or legal frame, some significant delineations emerge. Through a social or legal fame, two significant types of genetic group can be observed. For clarity, we will refer to the first type of group as a 'genetic class' and the second type of group as a 'genetic category'. The term 'genetic group' will be used as an umbrella term to refer to both.

Genetic classes map to easily recognised social groups. For example, ethnic groups or groups of sufferers of certain diseases – for example, sufferers of Down Syndrome.[3] Indeed, many genetic classes which may now be perceived through a genetic lens would claim to have existed long before advances in genetics allowed their genetic characterisation. These genetic classes will thus share the characteristics associated with other types of social, or even political, group. Members of such classes are likely to be aware of their status as a member of the class. They will be aware of what being a class member means to them, and to their lives and accordingly, may have particular desires which result from this membership. In turn, they are likely to know the other members of that class (or at least be able to get in con-

[3] http://www.downs-syndrome.org.uk/. (Last consulted 27.05.2015).

tact with, or find out about them). Through the communication of class members, a group identity may be established, as may the elaboration of common positions on matters of importance to the class. Certain such classes have found it useful to organise themselves and to establish communal decision making structures and channels of communication. Indeed, many such classes have found it useful to formalize aspects of the class in law – for example, representative organisations for certain ethnic minorities.

Genetic categories, however, have no independent social existence – although this may change with time as the significance of certain types of genetic architecture comes into clearer focus. For example, certain individuals may possess genetic architecture associated with predispositions – for example to disease or types of behaviour. Genetic categories have much in common with the algorithmic groups discussed elsewhere in this volume. Members of genetic categories may have no idea that they possess the relevant architecture placing them in a genetic category. They will unlikely be aware of other category members – not least as these other members may themselves be unaware of their category membership. In turn, category members may not have felt any impacts from their possession of the architecture in question. They may therefore have no sense of the consequence of their membership of the category. Without the ability to understand one's own category membership and the significance of this membership, and without the ability to share this experience with others, the genetic category will lack collective personality, identity or opinion. Such categories will lack organisational, decision making or communications structures.

It has long been established that the extraction and use of genetic information may pose a significant risk to the rights and interests of the individual from whom the biological sample was extracted (European Court of Human Rights 2008, §104). However data processing does not always need to focus on the individual. Indeed, in many cases, it is the group to which the individual belongs which is the focus of processing. When genetic groups are the focus of processing, a separate set of interests might be argued to be relevant alongside individual interests – those relating to groups themselves. Accordingly, the group might be recognised as a separate subject of legal protection.

10.4 Genetic Groups, Legitimate Interests and Legal Protection

In relation to genetic classes, an individual's identity is tightly tied up with the social groups that individual is part of. Accordingly, the autonomous existence of such groups is essential for the individual to be able to freely develop his or her personality in association with others. In turn, such groups play important social and political roles in society – indeed they are essential for the pluralism at the core of a democratic society (European Court of Human Rights 2005, §100). With this, these groups can be regarded to occupy a social space between the individual on the one hand, and society as a whole on the other. Such groups might thus be regarded, to a certain extent, as 'autonomous units' within society (Raab 2012). Indeed, they may be seen to have histories, identities, cultures and intentions which extend even beyond their temporary membership. The social existence of genetic classes places them within this category. With the recognition that such groups might be entitled to a degree of autonomy, a parallel recognition that they may have a claim to a degree of self-determination emerges (Laurie 2002). If data processing can impact an individual's right to self-determination, or can harm an individual, then comparable claims might be made on behalf of genetic classes.

Accordingly, it follows that genetic classes might have some claim over how data relating to them is processed. There is a small, but growing, body of legal and jurisprudential recognition for this position. Article 10 of the UNESCO Universal Declaration on the Human Genome and Human Rights states that: 'No research or research applications concerning the human genome… should prevail over respect for the human rights, fundamental freedoms and human dignity of individuals or, where applicable, of groups of people' (UNESCO 1997, §10). The Article 29 Working Party state that developments in the understanding of genetics may mean a 'legally relevant social group can be said to have come into existence – namely, the biological group' (Article 29 Data Protection Working Party 2004). In the *Havasupai* case, the Havasupai Indian Tribe argued that the class possessed dignitary interests which could be impacted through certain unconsented-to uses of members of the group's genetic information (Van Assche et al. 2013).

It is harder to justify a claim that genetic categories can have interests. If individuals do not know they are members of a category, how can a category be seen to have any desires, how can it be harmed – what would its interests be? Nevertheless, each individual member of a genetic category has an interest in the effects of genetic data processing on him or her. When genetic data relating to a genetic category is processed, each individual member of that category is thus potentially affected (European Court of Human Rights 2008, §104). On the one hand, this concern could be boiled down to a set of individual interests in not being harmed based on category membership. On the other hand, it has been observed that this is a highly 'atomistic ontology' and that the phenomena of processing on the level of categories of individuals might not be possible to deal with at the level of the single individual (Floridi 2014). An expedient response might be to recognise a collective interest on the level of the genetic category. The members of the category, taken together, can be seen to have a collective interest in how data relating to the category is processed and how that category is treated. As the category cannot itself communicate, recognising genetic categories as holding interests could be compared to recognising an incapax as holding interests (Beyleveld and Brownsword 2007). Thus, responsibility for category protection would always need to be delegated to an external party.

The fear that individual rights could be adversely affected through the use of genetic data related to groups has been reflected in certain genetic non-discrimination legislation. For example, the Council of Europe's Additional Protocol to the

Convention on Human Rights and Biomedicine, concerning Genetic Testing for Health Purposes states, in Article 4, that: 'Any form of discrimination against a person, either as an individual or as a member of a group on grounds of his or her genetic heritage is prohibited' (Council of Europe 2005, §4).

Whilst the idea of genetic groups as interest holders might seem odd, it should be recalled that the act of grouping and judging individuals according to perceived biological characteristics has a long and highly controversial history. The example which comes most prominently to mind is the horrific treatment of Jews, and others regarded as of inferior biology, by the Nazi regime in Germany. However, eugenics programmes, in one form or another, existed prior to the Nazis and have existed since. Indeed, the painful legacy of such ideas remains in modern society. For example, accusations of eugenics are often brought forward in the heated debates about pre-natal screening programmes and the legitimacy of terminating 'abnormal' pregnancies.[4] Learning from the lessons of the past, is it so strange that society should have rules related to acts dealing with such groups?

10.5 Data Protection Law and the Forthcoming Data Protection Regulation

Data protection law is the area of law outlining when data may be processed and under which conditions.

The current piece of legislation elaborating European data protection law is Directive 95/46 (European Parliament and European Council 1995). However, owing to changes in the technological and legal background to the Directive, its relevance and suitability came increasingly into question. Accordingly, 3 years ago, a process of reform of data protection law was started and in January 2012, the Commission released the proposed General Data Protection Regulation – the replacement to Directive 95/46 (European Commission 2012). The legislative process is now complete. The Regulation was adopted in 2016 and will enter into force in 2018.

The Regulation's scope extends to processing done in a wide range of contexts and by a wide range of actors. However, it only applies when the 'personal data' of a 'data subject' is processed. The Regulation specifically lists genetic data as personal data. Indeed the Regulation specifically defines genetic data as 'sensitive data', thereby making its processing subject to a stricter regime of protection. Whilst the Regulation does not apply to anonymous data, we believe – and have argued elsewhere – that it will rarely make sense to consider genetic data as anonymised (Hallinan et al. 2013).[5] Accordingly, genetic data should always be regarded as 'per-

[4] For a full discussion of eugenics to the present day see: Bashford and Levine 2010.

[5] We argue that, due to the uniquely individual nature of each genome, it is very difficult to claim any genetic data is anonymous.

sonal data'. Therefore, whenever genetic data are processed, the Regulation will apply.

The aim of the Regulation is dual. First, in each instance of data processing, there are legitimate rights and interests which should be taken into account. The Regulation aims to provide a procedure through which these interests can be taken into account and any conflict of interests can be resolved (De Hert 2009). On the one hand, individuals have rights, such as the right to privacy and to not be unfairly discriminated against, which can be affected by data processing. The Regulation aims to function as a system to make processing transparent to data subjects and to give them certain rights over their data. On the other hand, third parties – such as genetic researchers – need to process personal data in the pursuit of their own legitimate interests and these may, on occasion, override individuals' interests. Second, the Regulation aims to harmonize applicable data protection law in the EU.[6]

In relation to genetic data, the Regulation can be seen to employ four mechanisms which, together, aim to provide a procedure through which interests can be adequately taken into account and interest conflicts can be resolved.

1. Proposed processing of genetic data should be subject to an impact assessment, declared to, and checked in advance by, a supervisory authority.
2. Whenever personal data are processed, the data controller is subject to procedural, technical and organisational rules as to when and how data might be processed.
3. The data subject is granted a set of rights in relation to their personal data – including (in certain situations) the right to be asked for consent and a set of rights revolving around being informed about the processing of their data.
4. Finally, the Regulation provides for independent oversight of whether the provisions have been followed and for complaint and redress when they have not.

If genetic groups are seen to have legitimate interests to be protected, then data protection law would be a logical area of law to consider to carry the legislative burden of balancing these interests against other legitimate interests.

However, the idea of genetic groups as a focus of data protection law is novel. The Regulation – as with its predecessor the Directive – was drafted on the presumption that the individual, and individual rights, were the primary target of protection. Although data protection may seem like a – if not the – logical legal area of law through which to protect genetic groups' interests in the processing of genetic data, this is no guarantee that the Regulation will be a suitable instrument for this purpose.

In answering the question of suitability, the first question must be: Under which of the above mechanisms could genetic groups be included and under which would they be excluded?

[6] A Regulation is a specific type of legal instrument which is directly applicable in all EU states. This instrument is to be opposed to a Directive. Directives require transposition into national law. In the case of the Data Protection Directive, the divergence in national transpositions caused significant legal fragmentation.

10.6 1st Data Protection Mechanism: Advance Checking

As genetic data are regarded as 'sensitive' in the Regulation, they have been confirmed as data whose processing poses a particularly high risk to fundamental rights (Article 29 Data Protection Working Party 2011). Advance checking mechanisms are conceived of as legitimacy and proportionality controls and as mechanisms which allow harm mitigation in advance. Accordingly, when processing which may be particularly harmful is proposed, advance checking mechanisms become increasingly important and this processing is subject to increased scrutiny.

There are two forms of advance checking mechanism which are prominent in current data protection legal thought and which are required by the Regulation when sensitive data are processed. The first, and the more recent, is the Data Protection Impact Assessment (DPIA). The second is the obligation to consult with, and receive prior authorisation from, the appropriate Data Protection Authority.

The DPIA is a structured instrument an organisation planning to process personal data may use internally to analyse a potential processing operation. The instrument allows consideration of the proposed operation both in terms of its legitimacy under data protection law and in light of the broader harms it may cause. On the basis of the analysis conducted in the DPIA, the organisation might then take measures to address any identified issues. There is, in principle, no reason that the consideration of impact in a DPIA must be limited to one type of impact, or impact on one type of rights holder. In fact, some of the latest impact assessment methodologies with relevance for data processing have specifically included a consideration of both individuals and groups in their approach (Wright et al. 2014).[7] If genetic groups, of either type, can be seen to have legitimate interests in the processing of data, there is no reason that impacts on them could not be taken into account in a DPIA.[8]

The prior consultation and authorisation mechanism already exists in the Directive and is maintained in the Regulation. In this mechanism, the Data Protection Authority (DPA) is informed of the proposed processing. The DPA may then make a judgment as to whether, and if so under which conditions, the proposed processing might go ahead. This engages the Data Protection Authority as the body responsible for the application and enforcement of data protection law – i.e. they will evaluate the proposed processing in light of the requirements of data protection law. It also engages the DPA as the body responsible for more general oversight of data processing and the harms which might stem from data processing – i.e. as a proportionality mechanism. Accordingly, prior consultation and authorisation, by necessity, includes discretion on the part of Data Protection Authorities to deal with novel phenomena arising around data processing. The interests of genetic groups repre-

[7] See for example the methodology outlined by the SAPIENT Project: Wright et al. 2014.

[8] The Regulation recognises the necessity to take a broad approach to impacts, and to potentially affected parties, when conducting a DPIA. For example, Article 35(1) elaborates a DPIA is necessary when processing is 'likely to result in a high risk to the rights and freedoms of natural persons' – not just to data subjects.

sent such a phenomenon. Whilst they have not done so up to now, there is no reason that Data Protection Authorities could not use this discretion to take the interests of genetic groups, of either type, into account when consulting on and authorising processing operations.[9]

10.7 2nd Data Protection Mechanism: Organisational, Technical and Procedural Obligations

The Regulation lays out a number of obligations on the controller related to the conditions under which personal data might be processed. Broadly speaking, these obligations can be sorted into two types. First, data controllers are obliged to follow certain data processing principles. Second, data controllers are obliged to engage in data processing management practises.

The data processing principles which must be followed have a long history, stretching back to OECD privacy principles first outlined in 1980. Indeed, since then, the principles have been subject to only relatively minor change. The Directive outlined 5 main principles, each of which has been retained in the Regulation. Data must be[10]:

1. processed lawfully, fairly and in a transparent manner (to the data subject)
2. collected for specified, explicit and legitimate purposes
3. adequate, relevant, and limited to the minimum necessary in relation to the purposes for which they are processed
4. kept accurate and kept up to date
5. kept in a form which permits identification of data subjects for no longer than is necessary for the purposes [of processing].

There are only very few of these principles whose application is restricted to one, or one type of interested party (see principle 1, for example). The majority of these principles place obligations on the data controller without linking them to any specific concept of rights holder. There is thus no obstruction to including genetic groups, of either type, as the focus of these obligations. For example, there is no reason that data could not be kept accurate and up to date in relation to both individuals to whom they relate and to the genetic groups to whom they relate.

---

[9] The Regulation states that processing operations which result in the identification of a 'high risk' by a DPIA and whose risk has not subsequently been mitigated are also subject to the relevant Data Protection Authority's (DPA) prior authorisation and consultation – Article 36. The purpose of the impact assessment procedure is thus partially to provide relevant information for the supervisory authority related to the risks implied in an act of processing. If genetic groups were to be recognised as important in the assessment phase, this would mean they would also be important for a DPA when conducting a prior check and granting authorisation.

[10] These obligations are predominantly laid out in Article 5

Data processing management obligations aim to ensure that interests which can be affected by data processing are taken into account at each significant phase of a processing operation.[11] This includes consideration of interests in the phases prior to those involving actual data processing – for example, in the design phase with the principle 'data protection by design and default' – as well as in the systems – organisational and technical – which support data processing. Once again, data processing management principles are not linked to one type of rights holder. There is no reason, for example, that appropriate technical and organisational security measures could not be designed taking into account that a processing operation could pose a risk to both individuals and genetic groups.

10.8 3rd Data Protection Mechanism: Data Subject Rights

In the Regulation, the key parameter describing the legislator's concept of relevant rights holder is the concept of the 'data subject'. The 'data subject' concept, despite some criticism, has been retained from the Directive and is central to the framework outlined by the Regulation. The data subject has a set of rights relating to control over their data – for example, the right to consent to data processing. They also have a set of rights relating to transparency – for example, the right to be given information about processing. If a person or entity cannot qualify as a 'data subject' they cannot hold any of these rights.

The decisive qualification of the concept of 'data subject' in relation to genetic groups, is that a 'data subject' must be a 'natural person'.[12] Whilst it is true that the legal incorporation of a genetic class may qualify as a legal person, it is impossible for any group to qualify as a natural person.

The situation under the Regulation is thus cut and dried. However, a small amendment to the concept of 'data subject' would allow the inclusion of a broader range of entities as rights holders. The inclusion of genetic categories as rights holders – as 'data subjects' – is a non-starter. The concept of the 'data subject' is bound up with rights which require self-awareness and the ability to understand and communicate. However, the idea of giving genetic classes the status of rights holders deserves greater consideration. Indeed, there is historical precedent for the recognition of organised groups as 'data subjects'. Bygrave comments on the lack of distinction very early data protection laws drew between incorporated collective entities such as businesses and NGOs, and individuals (Bygrave 2002).

[11] These obligations are predominantly laid out in Chapter IV

[12] For clarification of the concepts of 'data subject' and 'personal data', see: Article 29 Data Protection Working Party 2007. See also Article 4(1) of the Regulation

10.9 4th Data Protection Mechanism: Oversight, Complaint and Redress

Oversight, complaint and redress mechanisms require the presence of something to oversee, to complain about and to redress. They do not exist independently of other provisions in the Regulation. In relation to these mechanisms, three questions might be asked: (1) What can be overseen and investigated? (2) Who can complain? (3) About what can complaints be made?

The DPA has the responsibility to oversee compliance with the Regulation as well as to investigate complaints. If genetic groups were to be recognised as holders of legitimate interests and to be entitled to certain protection under data protection law – for example, that should be considered in an assessment procedure or as entities to which obligations should be owed – the DPA could investigate whether obligations had been adequately observed. In data protection law, the power of a DPA to oversee and investigate has never been dependant on the registering of a complaint. This was the case in the Directive and remains the case in the Regulation. This is particularly significant in relation to the protection of genetic categories. The lack of communicative ability means that genetic categories are unlikely to be in a position to protect their own interests and to lodge complaints. Accordingly, to make sure any obligations owed to them were being properly observed, an independent body tasked with their protection would be necessary. With independent oversight and investigation power, the Data Protection Authority is in the position to fulfil this function.[13]

Once it has been established that group interests are to be protected, it is a short step to conclude that relevant groups should also have standing to lodge complaints with a DPA and beyond this, a right to a judicial remedy. The Regulation already foresees two circumstances in which groups could complain to a DPA or seek a judicial remedy: first, when legally constituted organisations have been mandated by a data subject; second, if Member State legislation allow legally constituted organisations to complain without the mandate of a data subject.[14] Accordingly, in the case of genetic classes, complaints could already be lodged on behalf of the class by a legally constituted representative body mandated by any member of that class who qualified as a data subject. In the case of a genetic category, provided a single data subject could be found, or a Member State exception was present, complaints could be lodged by organisations which are not constituted by the group itself, but which have a relevant interest in the category receiving protection. For example, it would not be unimaginable that an NGO dealing with genetic discrimination could lodge a complaint on behalf of a genetic category.

---

[13] The Regulation outlines the power to monitor and enforce data protection and to investigate on a Data Protection Authority's own volition, or on the basis of a complaint. See Article 58.

[14] See Article 80 of the Regulation. In relation to ensuring that the provisions of the Regulation are followed and that the oversight and complaint function has an impact, Article 58 outlines broad sanctioning powers for the DPA. Article 83 then set out that breaches of the Regulation are associated with potentially heavy financially penalties which can be levied by Data Protection Authorities.

Currently, the Regulation limits the subject of the complaint and redress mechanisms to violations of 'data subject' rights. With this language, complaints could not be made relating to any form of 'group' interest detached from the interests of individual data subjects. This could be easily changed – and would need to be – should relevant mechanisms be recognised to also apply to genetic groups. However, in the Regulation, there is no obstacle to a group lodging a complaint or judicial process on behalf of multiple data subjects simultaneously.[15] On the one hand, such an approach is not a substitute for the recognition of 'group rights' – the group itself, as opposed to its constituent members, is not the subject of the action. On the other hand, however, such an approach could allow a form of collective remedy to be sought – although there are still questions as to when and how such approach might be used (De Hert and Galetta 2015). Such an approach could, in some cases, function as a proxy for the direct protection of group interests.

In summary, there are few obstacles in the Regulation to the inclusion of genetic groups as subjects of protection. The one notable exception being the restrictive concept of 'data subject' which excludes all but 'natural persons'. However, with a small amendment, this obstacle could be cleared.

However, just because the Regulation has few impediments to including genetic groups as subjects of protection, does not mean that they should automatically be included. Accordingly, a second question might be asked: Would there be any problems associated with including genetic groups as a focus of protection?

10.10 Problems with Extending the Data Protection Regulation to Genetic Groups

In the first instance, we see three legal technical issues to using the Regulation as a framework for the protection of genetic groups. On top of this, we would like to point out a compelling substantial concern regarding genetic groups as being owed obligations, or as being rights holders, under the Regulation. Despite the apparent severity of the objections we point out, we do see a way forward, and this will be elaborated in the section following this one.

First, for legislation to be practically functional, it requires clarity *rationae personae*. In relation to the individual, concepts of 'data subject' and 'personal data' have served to provide this clarity. These clarify who the subject of protection is, and when there is sufficient link between them and data processing to engage their interests and qualify them for protection. To make the Regulation functional in relation to genetic groups, a clear *rationae personae* for these groups would thus also be necessary. For example, in relation to the effective conduct of a DPIA which took groups into account, the controller would need to know which genetic groups to consider.

[15] See Article 80

However, there are a huge number of genetic groups which could be recognised in any single genetic sample. Indeed, this number is subject to fluctuation as genetic science develops and human characteristics are found to be (or not to be) genetically influenced. Furthermore, as discussed above, each genetic group is nested within a number of other possible genetic groups. Eventually, any genetic information extracted from an individual could be used to inform judgments about all other humans, and in turn all genetic groups (Juengst 1998). It would be impossible for a data controller to take all groups into account. The link between an individual data subject and their personal data was established on the basis whether the data could identify him or her.[16] Such an approach would be irrelevant in relation to genetic groups. Alternative guidelines would be needed to establish inclusion and exclusion. Although some research projects have sought to develop methodologies which take algorithmic groups into account – for example, the SAPIENT project considered groups constructed by smart surveillance algorithms – there are currently no guidelines available which coherently identify relevant and non-relevant genetic groups (Wright et al. 2014).

Second, each mechanism in the Regulation is populated by concepts and definitions. Most of these have been taken on from the Directive and accordingly, their meaning and application has already been refined through time and use. However, this refinement has been done in relation to the concept of the individual as the relevant interest holder. It is not a certainty that these interpretations will remain relevant when the interest holder is a genetic group.

We might take the principle that data should be accurate and kept up to date as an example. It has been regarded as relatively easy to apply in relation to individual personal data. The Information Commissioner's Office in the UK links accuracy to the representation data makes of a fact that either is, or isn't, true for an individual. They state: 'It will usually be obvious whether information is accurate or not'.[17] But the idea that such facts are easily definable in relation to genetic groups, or that they will be easily recognisable as true or false is questionable. What would accuracy mean in relation to genetic group data? Would it be a reference to the accuracy of the original data sets collected from individual group members, or would it refer to results of judgments referring to the group as a whole, or perhaps both? In turn, what would it mean to keep data about a group 'up to date'? The consistency of a genetic groups is likely to change over time. For example, some members will die, others will be born. Equally, there will be scientific changes in the understanding of the nature of a certain genetic group – for example the association of two previously unrelated areas of the genome as relevant to a disease. This will also serve to change the consistency of the group. Would keeping data about a group up to date mean adding and subtracting samples and genomic data to reflect changing understanding of the group?

---

[16] For a discussion of identifiability and how data must relate to a person, see: Article 29 Data Protection Working Party 2007.

[17] See: https://ico.org.uk/for-organisations/guide-to-data-protection/principle-4-accuracy/. (Last consulted 27.05.2015).

Finally, we have observed that genetic groups could be owed obligations (in Sect. 10.7) and that genetic classes could be included as rights holders with a small alteration in the concept of 'data subject' (in Sect. 10.8). However, data protection law, as outlined in the Directive and the Regulation, has aimed to resolve the interest conflicts in data processing between one data subject and one data controller. It is around this bilateral relationship that the allocation of rights and obligations in the Regulation has been shaped.

Recognising genetic groups as subjects of protection – either in terms of being owed obligations or as being rights holders – would create a new dynamic. There would now be three types of interest to take into account in any processing operation. Indeed, as genetic groups are nested within one another, it is not hard to imagine that processing operations could throw up multiple genetic groups with interests to be taken into account (Greely 2001). This would create new forms of interest conflict. The original conflict of interests between controller and individual 'data subject' would remain but three new forms of conflicting interests would also come into view: first, between 'individual data subject' and genetic group(s); second, between data controller and genetic group(s); third, between relevant genetic groups. From a legal technical perspective, the current Regulation lacks the complexity to deal with this new dynamic. For example, let us imagine that the concept of 'data subject' were to be extended to genetic classes. Under the current Regulation, consent would thus presumably be required from both the source individual, and from any relevant genetic class. If all agreed that processing should go ahead, there would be no problem. But if the individual gave consent, but the class refused (or one class refused and another consented) there would need to be a process laid out for deciding how to move forward (Taylor 2012). I.e. a process for deciding whose interest should be overridden. The Regulation does not provide this.

Following from the above legal technical obstacles, there is a compelling substantial objection to simply dropping genetic groups into the Regulation's current system of rights and obligations.

The balance between the individual and the data controller struck by the Regulation is the result of decades of legislative evolution. Within this period, the consequences of different types of processing technologies have been carefully observed and considered, as have the societal principles which they may affect and the competing interests they may serve (Kosta 2013). If genetic groups were simply to be added as a subject of protection without further consideration, they would occupy a role comparable to that the individual currently occupies. The data controller would thus need to discharge obligations to multiple parties. This might impose a heavy burden – financially and organisationally – on the data controller. Where data subject rights were engaged, the data subject would be put in a position where the possibility to enjoy their rights could be obstructed by the genetic groups to which they belong – the tyranny of the genetic group would be a real possibility.

This seems an undesirable scenario. There remain few sectors – medical research and in exceptional cases in law enforcement – in which genetic data is routinely used to make judgments about groups. There is no European jurisprudence related to genetic groups (or indeed, with certain notable exceptions, much related to the processing of genetic data in general). Consideration of the issue of group rights in genetics remains – in Europe at least – largely theoretical. In short, there is currently little evidence to suggest that genetic groups' interests should be valued in the same way as the individual's (Taylor 2012). Just because an interest may be recognised, does not mean that it is equivalent to all other interests.

In summary, the Data Protection Regulation provides the facility for comprehensive protection for both types of genetic group. However, a blunt and unconsidered extension of protection is likely to cause a number of problems. Both the processing of genetic data and the issue of group protection are novel phenomena. Proportionate solutions to such novel phenomena are unlikely to arise from unsubtle top-down approaches.

10.11 Moving Forward

Given that it would be problematic to simply extend the same protection to groups as enjoyed by individuals, alternative approaches must be considered. The first option would be to do nothing and to maintain the current status quo. Considering the lack of jurisprudential consideration the issue of the genetic groups has received, doing nothing would not be catastrophic. However, we believe that the arguments which have been put forward about genetic groups having interests are strong enough to be taken seriously. This is particularly true in light of the history of genetic profiling. Doing nothing would be equivalent to ignoring these arguments and this would be short-sighted. Accordingly, we believe a second approach – a middle-way – to be preferable. This approach offers a working solution to the first technical problem we observed. It then offers an approach through which the other two technical problems, and the substantial problem, can be engaged with – if not immediately solved.

There has been an explosion in the scale of processing of genetic data in the last few years. Despite this, there has been relatively little consideration of the issue of genetic groups by bodies responsible for guidance and interpretation of data protection law. The Article 29 Working Party last considered the issue of genetic data in 2004. In this consideration, they specifically mentioned the communal nature of genetic data – for example in their recognition that a 'legally relevant social group can be said to have come into existence – namely, the biological group'. A first step is thus to re-raise awareness of the issues connected with the processing of genetic data as well as the communal nature of this data. Accordingly, it would be advisable, for the European Data Protection Board – as the successor to the Article 29 Working Party, outlined by the Regulation – to revisit the issue.[18]

The first thing such guidance would need to address is the issue of inclusion and exclusion of genetic groups as subjects of consideration in data processing. As we

[18] See Article 68 of the Regulation.

have already clarified, it is not relevant to use approaches to individual *rationae personae* in relation to genetic groups. Not least, as genetic groups are nested within countless other genetic groups. A tentative alternative to a content-based approach might be to focus on the aim of processing. For example, if research was being done on mutations in the HEXA gene, then those with that mutation could be regarded as the relevant group, and all other genetic groups could be excluded as non-relevant.

Following a clarification of which groups might be relevant in relation to a processing operation, guidance could outline an approach to protection. Guidance should take care to avoid the technical and substantive problems listed in the previous section associated with genetic groups being owed obligations or being rights holders. This can be done by initially focussing on the *ex ante* checking and control mechanisms. As an impact assessment needs to be conducted whenever genetic data are processed, the additional consideration of whether genetic groups are a target of processing in an assessment would not place a large burden on data controllers and would have no impact on data subjects. In turn, the assessment procedure is not legally binding, nor is it attached to owing genetic groups specific obligations or granting them rights.

If guidance were to elaborate the inclusion of impacts on genetic groups as subject matter for DPIAs, information would be generated related to groups in each case in which genetic data were to be processed. On the basis of this information, the responsible local DPA would thus be in the position to consider each case on its merits. In some cases, genetic groups might need protection; in other cases this may not be wise. In each case, the Data Protection Authority might use their discretion as to whether, how, and which data protection mechanisms might apply. This discretion would allow a bespoke adaptation of the meaning of concepts and definitions in specific, concrete instances – for example, a clarification of the concept of 'accuracy'. It would also allow a mix-and-match approach to the application of mechanisms (or aspects of mechanisms) so as not to disproportionately interfere with the existing rights of controllers or data subjects.

Over time, DPAs consideration of cases may create jurisprudence around when and how genetic groups should be protected through data protection law. Out of this bottom-up jurisprudence, more general principles may slowly be distilled.

10.12 Conclusion

In this contribution it has been observed that individuals might be grouped according to shared genetics and that these groups might be seen to have legitimate interests engaged when data related to them is processed. Two types of genetic group were identified, the genetic class – which maps to an already recognised social group – and the genetic category – a genetic group identifiable only through members' shared genetics.

If genetic groups are seen to have legitimate interests to be protected, then data protection is one area of law we might look at to carry some of the legislative burden of balancing these interests against other legitimate interests. In the future, the Data Protection Regulation will be the key instrument elaborating data protection law. Accordingly, this contribution has considered: first, whether the Regulation has the facility to accommodate the protection of genetic groups; second, what consequences including genetic groups as subjects of protection would have; third, based on the opportunities and problems isolated, what a proportionate approach moving forward might look like.

It was observed that the Regulation employs four mechanisms to protect interests which might be impacted by processing

1. Advance checking
2. Data controller obligations
3. Data subject rights
4. *Ex post* checking and recourse

The contribution first considered how far both types of genetic group could be considered as subjects of protection under each mechanism. Interestingly – especially considering how much has been made of the highly individualistic focus of the Regulation – many of the mechanisms could easily be extended to protect both forms of genetic group.

The advance checking mechanisms – DPIA and DPA consultation and authorisation – are flexible, open ended mechanisms. They do not need to be tied to a specific type of rights holder. With the relevant consideration, these could easily be extended to include consideration of both genetic classes and genetic categories.

The same is true for data controller obligations. Only a limited number of these must be owed to one, or one type of, rights holder. The obligation to hold data securely, for example, could as easily be owed to an individual data subject, relevant genetic classes and relevant genetic categories. Indeed, there is no reason it might not be owed to all simultaneously.

Giving genetic groups rights, on the other hand, is somewhat more complicated. In the first instance, genetic categories have no ability to communicate and therefore are not subjects to whom actionable rights can be given. Although genetic classes may constitute potentially rights-holding entities, the Regulation continues to rely on the concept of data subject to define rights holders. As only natural persons can be data subjects, even genetic classes are excluded.

Finally, *ex post* checking and recourse mechanisms become relevant only when there is something to check and to claim recourse in relation to. To rely on these mechanisms, recognition for groups' interests and how data protection law served to protect them would already need to have been clarified. However, should this happen, there is no objection to including either type of groups' interests as the subject of protection. In particular, the ability of the DPA to investigate without a complaint being filed could serve to protect both genetic classes and genetic categories.

However, simply because the Regulation can apply to genetic groups, does not mean it should. The contribution considered the consequences of including genetic groups as subjects of protection and revealed a number of awkward problems.

First, the concept of the scope of the Regulation *materiae personae* as it applies to the individual, does not easily translate to the genetic group. In any instance of genetic data processing, countless groups might be recognized as relevant. New criteria would be required to clarify what a genetic group is, and which groups to include and exclude as subjects of protection.

Second, the definitions and concepts populating the Regulation have been developed with the individual in mind and may also need some clarification to be applicable to genetic groups – for example, what would the principle of data accuracy mean in relation to data about a genetic group?

Third, the set of rights and obligations laid out by the Regulation have been designed as a process to resolve a bilateral interest conflict. Including genetic groups would throw up a multilateral interest conflict. The Regulation lacks the facility to reconcile such a conflict.

Finally, we observe that including genetic groups in the current system of rights and obligations without further consideration would be to falsely equate the value of the interests of the group and the individual data subject.

It would thus be problematic to simply apply the Regulation to genetic groups as well as individuals. However, doing nothing ignores that there are good arguments supporting the suggestion that genetic groups can have interests.

A middle ground is needed in which these arguments are taken seriously, but problems related to overregulation through a top-down approach are avoided. Accordingly, we propose an approach based on guidance and case-by-case consideration. This is a soft approach but a proportionate one.

A first step would be for the European Data Protection Board, to revisit the issue of genetic data, this time including a consideration of genetic groups. This guidance could raise awareness of the issue of genetic groups and provide an approach which could be followed Europe wide.

The guidance could begin by outlining when a genetic group might be regarded as having interests to be taken into account. We have suggested that the traditional approach to establishing a link between 'data subject' and 'personal data' may not be effective. Instead, perhaps an approach based on processing intention might be considered.

An approach to protection might then be outlined. We believe this approach should focus on the *ex ante* checking mechanisms – DPIA, Data protection Authority prior checking and consultation.

On the basis of information generated through the DPIA procedure, the relevant DPA would then have discretionary power to consider whether, which and how mechanisms should apply to protect groups in concrete cases. Over time, out of the jurisprudence created by DPAs' consideration of specific cases, more general principles might be distilled.

Bibliography

Article 29 Data Protection Working Party. 2004. *Working document on genetic data*, WP 91

Article 29 Data Protection Working Party. 2007. *Opinion 4/2007 on the concept of personal data*, WP136, 2007.

Article 29 Data Protection Working Party. 2011. *Advice paper on special categories of data ("sensitive data")*, Ares (2011) 444105.

Aubret, F., R. Shine, and X. Bonnet. 2004. Adaptive developmental plasticity in snakes. *Nature* 431: 261–262.

Bashford, A., and P. Levine (eds.). 2010. *The Oxford handbook of the history of eugencis*. Oxford: Oxford University Press.

Beisson, J. 2008. Performed cell structure and cell heredity. *Prion* 2(1): 1–8.

Beyleveld, D. 2004. An overview of directive 95/46/EC in relation to medical research. In *The data protection directive and medical research across Europe*, ed. D. Beyleveld et al., 5–23. Aldershot: Ashgate.

Beyleveld, D., and R. Brownsword. 2007. *Consent in the law*. Oxford: Hart.

Bygrave, L. 2002. *Data protection law: Approaching its rationale, logic and limits*. The Hague: Kluwer Law International.

Council of Europe. 2005. *Additional protocol to the convention on human rights and biomedicine, concerning genetic testing for health purposes*, CETS No. 195. Available at http://conventions.coe.int/Treaty/EN/Treaties/Html/195.htm

De Hert, P. 2009. *Citizens' data and technology: An optimistic perspective*. The Hague: Dutch Data Protection Authority.

De Hert, P., and A. Galetta. 2015. The proceduralisation of data protection remedies under EU data protection law: Towards a more effective and data subject-oriented remedial system. *Review of European Administrative Law*, Special Issue, 123–148.

European Commission. 2012. *Proposal for a regulation of the European Parliament and of the council on the protection of individuals with regard to the processing of personal data and on the free movement of such data (General Data Protection Regulation)*, COM(2012) 11 final.

European Parliament and Council, Regulation (EU) 2016/679 of the European Parliament and of the Council of 27 April 2016 on the protection of natural persons with regard to the processing of personaldata and on the free movement of such data, and repealing Directive 95/46/EC (General Data ProtectionRegulation), O.J., L 119/1

European Court of Human Rights. 2005. *Leyla Şahin v. Turkey*, no. 44774/98.

European Court of Human Rights. 2008. *S. and Marper v United Kingdom*, no. 30562/04 and 30566/04.

European Parliament and Council, Directive 95/46/EC of the European Parliament and of the Council of 24 October 1995 on the protection of individuals with regard to the processing of personal data and on the free movement of such data, OJ, L281/31, 1995

Floridi, L. 2014. Open data, data protection, and group privacy. *Philosophy and Technology* 27: 1–3.

Greely, H. 2001. Informed consent and other ethical issues in human population genetics'. *Annual Review Genetics* 35: 785–800.

Hallinan, D., M. Friedewald, and P. de Hert. 2013. Genetic data and the data protection regulation: Anonymity, multiple subjects and a prohibitionary logic regarding genetic data? *Computer Law & Security Review* 29(4): 317–329.

Hartl, D.F., and M. Ruvolo. 2012. *Genetics: Analysis of genes and genomes*, 8th ed. Burlington: Jones and Bartlett Publishing.

Juengst, E. 1998. Groups as gatekeepers to genomic research: Conceptually confusing, morally hazardous, and practically useless. *Kennedy Institute of Ethics Journal* 8: 183–200.

Kosta, E. 2013. *Consent in European data protection law*. Leiden: Brill.

Laurie, G. 2002. *Genetic privacy: A challenge to medico-legal norms*. Cambridge: Cambridge University Press.

Lowe, A., et al. 2001. Inferring ethnic origin by means of an STR profile. *Forensic Science International* 119: 17–22.

Nomper, A. 2005. *Open consent – A new form of informed consent for population genetic databases*, Doctor Iuris, Budapest/Oxford/Tallinn: 2005. Available at http://dspace.utlib.ee/dspace/bitstream/handle/10062/818/nomper.pdf?sequence=5

Nuffield Council of Bioethics. 2007. *The forensic use of bioinformation: Ethical issues*. Available at http://nuffieldbioethics.org/wp-content/uploads/The-forensic-use-of-bioinformation--ethical-issues.pdf

OECD. 1980. *OECD guidelines on the protection of privacy and transborder flows of personal data*. Updated version available at: http://www.oecd.org/sti/ieconomy/oecdguidelinesontheprotectionofprivacyandtransborderflowsofpersonaldata.htm

Raab, C. 2012. Privacy, social values and the public interest. In *Politik und die Regulierung von Information* ['Politics and the Regulation of Information'], Politische Vierteljahresschrift Sonderheft 46, eds. A. Busch, and J. Hofmann, 129–151. Baden-Baden: Nomos Verlagsgesellschaft.

Rees, J. 2003. Genetics of hair and skin color. *Annual Review Genetics* 37: 67–90.

Rouvroy, A. 2008. *Human genes and neoliberal governance: A foucauldian critique*. Abingdon: Routledge-Cavendish.

Taylor, M. 2012. *Genetic data and the law: A critical perspective on privacy protection*. Cambridge: Cambridge University Press.

UNESCO. 1997. *Declaration on the human genome and human rights*. Available at: http://portal.unesco.org/en/ev.php-URL_ID=13177&URL_DO=DO_TOPIC&URL_SECTION=201.html

Van Assche, K., S. Gutwirth, and S. Sterckx. 2013. Protecting dignitary interests of biobank research participants: Lessons from Havasupai tribe v Arizona board of regents. *Law, Innovation and Technology* 5(1): 54–84.

Wright, D., et al. 2014. *A guide to surveillance impact assessment*, D 4.4. Available at http://www.sapientproject.eu/D4.4%20-%20SIA%20Manual%20%28submitted%2001%20August%202014%29.pdf

Chapter 11
Do Groups Have a Right to Protect Their Group Interest in Privacy and Should They? Peeling the Onion of Rights and Interests Protected Under Article 8 ECHR

Bart van der Sloot

Abstract Privacy is perhaps the most elusive of all human rights – difficult to define, dependent for its meaning on context, epoch, person and culture and contested ever since it was first formulated. One of the reasons is that privacy is at the same time both the most individual and the most general, the most personal and the most abstract of all human rights. The right to privacy under the ECHR originates in the doctrine simply prohibiting states to abuse their powers. Consequently, a right to complain about the abuse of power was granted not only to individuals, but also to legal persons, groups and states, as the value at stake with privacy violations was a societal interest. Gradually, under the interpretation of the ECtHR, the right to privacy has become more and more focused on natural persons and individual interests, so that groups and legal persons are in principle denied a right to complain under Article 8 ECHR. This paradigm has functioned relatively well for decades as most privacy violations were targeted at specific individuals. However, under the current technological paradigm, often referred to as big data, the threats to privacy increasingly do not materialize on an individual level, but on a general or group level. Should groups then be allowed to invoke a right to privacy to protect their own interest?

Keywords Group privacy • Legal persons • Privacy • Data protection • Rights • Interests

B. van der Sloot (✉)
Tilburg Institute for Law, Technology and Society, Tilburg University, 90153, Warandelaan 2, 5000 LE Tilburg, The Netherlands
e-mail: B.vdrSloot@uvt.nl

L. Taylor et al. (eds.), *Group Privacy*, Philosophical Studies Series 126,
DOI 10.1007/978-3-319-46608-8_11

11.1 Introduction

The right to privacy under the European Convention on Human Rights (ECHR) originates in the doctrine simply prohibiting states to abuse their powers. States may use their powers for legitimate reasons and in doing so, they may limit the human rights of citizens. Inter alia, a state is qualified to enter the home of a specific person if it has reason to belief that the person has, for example, committed a murder. However, states may not use their powers to randomly and arbitrarily enter the homes of citizens, wire-tap telecommunications without a specific reason or execute body-cavity searches on the basis of race rather than an objective criterion. While this focus on the duty of states not to abuse their power (instead of subjective rights of natural persons to protect their own interests) is still an important pillar of Article 8 of the Convention, specifying the right to privacy, the European Court of Human Rights (ECtHR) has focused more and more on the individual and his interests. In this line of argumentation, the Court has stressed that privacy is the most personal of all human rights granted protection under the Convention. The right to privacy, accordingly, only protects a person's individual autonomy, human dignity and personal freedom. This contrasts with the freedom of expression, for example, which safeguards the individual interest of a person to express himself, but also protects societal interests, among others by facilitating the search for truth through the market place of ideas and safeguarding a free and independent press, conceived as a precondition for every modern democracy.

The difference between the focus on the duty of state not to abuse its powers and the focus on the rights of individuals to protect their own interests has led to a debate not only regarding the interests that are protected by the right to privacy, but also regarding the question of who has a right to complain about a violation of Article 8 ECHR. Under the first interpretation, the focus is not on the specific interests of the claimant, but on the actions of the state. The right to complain about these types is necessarily wider than if the focus is on individual interests of natural persons, which in principle can only be invoked and defended by the natural persons themselves. Consequently, while the drafters of the ECHR opened up the right to complain about the violation of the right to privacy by a state (note: the Convention only allows complaints about the conduct of states, not about the potential infringement on human rights by legal or natural persons) to natural persons, legal persons, groups and other states, the ECtHR, focussing on individual rights and individual interest, in principle only allows natural persons to invoke Article 8 ECHR. Only under exceptional circumstances is it willing to relax this narrow interpretation.

Article 8 ECHR specifies: '1. Everyone has the right to respect for his private and family life, his home and his correspondence. 2. There shall be no interference by a public authority with the exercise of this right except such as is in accordance with the law and is necessary in a democratic society in the interests of national security, public safety or the economic wellbeing of the country, for the prevention of disorder or crime, for the protection of health or morals, or for the protection of the rights and freedoms of others.' The ECHR contains two independent complaint procedures. Article 33, regarding inter-state complaints, specifies: 'Any High

Contracting Party [a state having signed onto the Convention] may refer to the Court any alleged breach of the provisions of the Convention and the Protocols thereto by another High Contracting Party.' Article 34, regarding individual complaints, holds: 'The Court may receive applications from any person, nongovernmental organisation or group of individuals claiming to be the victim of a violation by one of the High Contracting Parties of the rights set forth in the Convention or the Protocols thereto. The High Contracting Parties undertake not to hinder in any way the effective exercise of this right.' In consequence, the Convention allows states to submit a complaint before the Court and three types of individual complainants: natural persons, legal persons and groups.

It is especially the rights of groups that will be central to this contribution, but in order to understand their position in relation to the rights of other parties and the types of interests that are central to their claims, a brief overview will be provided of other types of interests and rights being put forward in relation to Article 8 ECHR. This contribution will answer two questions, one descriptive and one normative. First, which types of rights and interests are currently accepted under the scope of the right to privacy under the European Convention on Human Rights? It will answer this question by analysing the case law of the European Court of Human Rights and the former European Commission of Human Rights (ECmHR), which was abolished in 1998 when its tasks (to assess the admissibility of complaints) were transferred to a separate chamber of the Court. It will be argued that while most types of rights (those of natural persons, legal persons and states) and most types of interests (individual interests, those of legal persons and general interests) are recognized by the Court, the right of groups to protect their own interests is not or only marginally so. Second, should groups be allowed to claim a right to privacy? To answer this question, this chapter will discuss why there might be a need for such a right under the current technological paradigm, often referred to as big data, as well as which reasons exist for not granting a right to groups and in how far these objections can be overcome. Big data may be defined as gathering massive amounts of data without a pre-established goal or purpose, about an undefined number of people. These data are then processed on a group or aggregated level through the use of statistical correlations.

This contribution will take the four parties specified under the Convention as a starting point and analyse whether they are allowed to invoke a right to privacy under the Convention. It will also discuss which types of interests may be served with privacy protection under the ECHR. To simplify, this chapter will take four types of interests, those of individuals, those of legal persons, those of groups and those of the state, that is, society in general. Thus, while the first three are specific interests of individuals, legal persons or groups, the fourth is a general interest, although of course these may overlap. The constant stigmatization and discrimination with regard to minority groups, for example, may be regarded as a particular interest of that group, but may also have a real and concrete impact on society as a whole. The same counts for individual interests. Individuals belonging to a minority group have a particular and individual interest in not being stigmatized or discriminated, but this is part of the shared and common interest of that group. Likewise, with individuals owning or strongly connected to a legal person, for example, their own business, the interests of the individual and the legal person may be intertwined. Finally, it should be kept in mind that with inter-state complaints, states do complain about the general, societal interest of a nation, though this is not the interest of their own nation, but that of another country suspected of engaging in systemic human rights violation (for example, France might argue that Turkey is violating the basic human rights of its population). Although there are many difficulties with this division, it will still help to better distinguish between and understand the different types of interests, rights and rationales involved in privacy protection under the European Convention on Human Rights.

The table below shows the different rights and interests. Consequently, an individual may invoke his right to privacy in order to protect his own interests, but he can also rely on Article 8 ECHR, through a class action, to protect the general interest or the interests of a (minority) group. Likewise, a group may (theoretically) invoke the right to privacy to protect its own interests, but it may also do so to protect the interests of a specific individual or by relying on a general interest. Section 11.2 will discuss whether and if so, in how far states may rely on the right to privacy in order to protect individual interests, the interests of a legal person, the interests of a (minority) group and the general interest. Section 11.3 will analyse whether and if so, in how far individuals have a right to rely on Article 8 ECHR to protect their own interests, the interests of legal persons and/or those of (minority) groups. Section 11.4 will discuss the cases in which either individuals, legal persons or groups submit a complaint under Article 8 ECHR in order to protect the general interest: class actions. Section 11.5 will analyse whether and if so, in how far legal persons can rely on the right to privacy to protect the interests of individuals, their own interests and/or those of groups, and Sect. 11.6 will do the same for groups. Finally, Sect. 11.7 will provide an analysis and tackle the normative question: should groups be allowed to invoke a right to privacy in order to protect their own interests?

Interest				
Right	Individual	Legal person	Group	General
Individual	Section 3	Section 3	Section 3	Section 4
Legal person	Section 5	Section 5	Section 5	Section 4
Group	Section 6	Section 6	Section 6	Section 4
State	Section 2	Section 2	Section 2	Section 2

11.2 The State's Right to Invoke the Right to Privacy

When the Universal Declaration on Human Rights (UDHR) and in its wake the European Convention on Human Rights were drafted, the Second World War had just ended and while most fascist regimes had fallen, totalitarian regimes in Communist countries still thrived. The core vision behind both documents was consequently simply to prevent the abuse of power by states. The human rights violations that took place were not so much targeted at specific individuals, rather, large groups in society were denied their most basic rights and freedoms. This not only regarded groups such as Jews, gays and Gypsies, who were the targets of abusive practices, other human rights violations affected larger groups in society as well. For example, the problem with secret services such as the Stasi was not so much that the privacy of specific individuals was infringed, but rather that it collected data about everybody living in the Deutsche Demokratische Republik.[1]

Consequently, the Convention as a whole and the right to privacy in particular were focused on a general duty of the state not to abuse its powers. Of all articles contained in the Convention, the rationales of negative obligations for the state and negative freedom for individuals are most prominent in the right to privacy under Article 8 ECHR. Under the Declaration, on which this provision is based, it was this Article that was originally plainly titled 'Freedom from wrongful interference'.[2] Likewise under the Convention, the right to privacy was originally only concerned with negative liberty, contrasting with other qualified rights in which positive freedoms are implicit, such as a person's freedom to manifest his religion or beliefs (Article 9), the freedom of expression (Article 10) and the freedom of association with others (Article 11). Likewise, the wording of Article 8 ECHR does not contain any explicit positive obligation, such as, for example, under Article 2, the obligation to protect the right to life; under Article 5, to inform an arrested person of the reason for arrest and to bring him promptly before a judge; under Article 6, the obligation to ensure an impartial and effective judicial system; and under Article 3 of the First Protocol, the obligation to hold free elections.[3]

The fact that the Convention was designed against the background of the abusive practices during the Second World War of course affected the way in which the rights and freedoms contained in the ECHR were designed by the authors of the Convention. The drafters of the Convention also distinguished between the Commission and the Court. The Commission had no authority other than filtering cases; it could declare cases admissible or inadmissible for a variety of reasons. It did not, however, have the power to decide on the substance of the matter. The Court did. While individual applicants had a right to petition to the Commission (the current Article 34 ECHR), they did not have a similar right to take the case to the ECtHR. Only the Commission or a state could take a case to Court, even if an individual complainant (natural person, group or legal person) had originally submitted a case and even if it was declared admissible by the Commission.[4] States, of course, could also submit an inter-state complaint (the current Article 33 ECHR) – in this case, states also had direct access to the Court. Consequently, there are two procedures. First, states can submit an inter-state complaint – this is mostly in the general interest, for example if a country violates human rights on mass scale. Second, states (and the Commission) could pursue the claim of a specific individual, legal person or group. While individual complainants are currently also allowed to submit

[1] See for a further exploration: van der Sloot (2014).

[2] UN documents: E/HR/3.

[3] Tomlinson, H. (2012) p. 2.

[4] <http://www.echr.coe.int/Documents/Collection_Convention_1950_ENG.pdf>.

a complaint before the Court directly (see Sect. 11.3), and the system of inter-state complaints is very seldom practiced (van Dijk et al. 2006), the Convention still allows states to invoke the right to privacy and submit an inter-state complaint in order to protect either the interests of specific individuals, legal persons or groups or the general interest.

Interest				
Right	Individual	Legal person	Group	General
Individual				
Legal person				
Group				
State	x	x	x	x

11.3 Individuals' Right to Invoke the Right to Privacy

Over time, the Convention has been revised on a number of points, so that, inter alia, individual complainants (individuals, groups and legal persons) have direct access to the Court to complain about a violation of their privacy (the task of the Commission being reassigned to a separate chamber of the Court – the two-tiered system still exists).[5] The Court has also made some major steps to revise the meaning and interpretation of the right to privacy under the Convention. Among other matters, it has accepted that Article 8 ECHR not only protects the negative freedom of citizens, but also the right to develop one's personality to the fullest, and has stressed that states may not only have a negative duty not to abuse its powers, but also a positive duty to use its powers to protect its citizens and to facilitate their quest for full personal development.[6] Moreover, the Court has placed a very large emphasis on individual interests and personal harm if it assesses a case regarding a potential violation of Article 8 ECHR. This focus on individual harm and individual interests brings with it that certain types of complaints are declared inadmissible by the European Court of Human Rights, which means that the cases will not be dealt with in substance.[7]

So called *in abstracto* claims are in principle declared inadmissible. These are claims that regard the mere existence of a law or a policy, without them having any concrete or practical effect on the claimant.[8] *A-priori* claims are rejected as well, as the Court will usually only receive complaints about injury which has already mate-

[5] Protocols and 11 to the Convention.

[6] B. van der Sloot, 'Privacy as human flourishing: could a shift towards virtue ethics strengthen privacy protection in the age of Big Data?', JIPITEC, 2014–3.

[7] See about the focus on individual rights and individual interests with respect to data protection: B. van der Sloot, 'Do data protection rules protect the individual and should they? An assessment of the proposed General Data Protection Regulation', International Data Privacy Law, 2014–4.

[8] ECtHR, Lawlor/UK, appl.no. 12763/87, 14/07/1988.

rialized. Claims about future damage will in principle not be considered.[9] Hypothetical claims regard damage which might have materialized, but about which the claimant is unsure. The Court usually rejects such claims because it is unwilling to provide a ruling on the basis of presumed facts. The applicant must be able to substantiate his claim with concrete facts, not with beliefs and suppositions. The ECtHR will also not receive an *actio popularis,* a case brought up by a claimant or a group of claimants, not to protect their own interests, but that of others or society as a whole. These types of cases are better known as class actions.[10] Then there is the material scope of the right to privacy, Article 8 ECHR. In principle, it only protects the private life, family life, correspondence and home of an applicant. However, the Court has been willing to give a broader interpretation, for example, it has held that it also protects the personal development of an individual, it includes protection from environmental pollution and may extent to data protection issues. Still, what distinguishes the right to privacy from other rights under the Convention, such as the freedom of expression, is that it only provides protection to individual interests.[11]

This focus on individual interests has also had an important effect on the types of applicants that are able to submit a complaint about the right to privacy. Although the Court has accepted that churches may invoke the freedom of religion (Article 9 ECHR) and that press organizations may rely on the freedom of expression (Article 10 ECHR), it has said that in principle, only natural persons can invoke a right to privacy. For example, when a church complained about a violation of its privacy by the police in relation to criminal proceedings, the Commission found that '[t]he extent to which a non-governmental organization can invoke such a right must be determined in the light of the specific nature of this right. It is true that under Article 9 of the Convention a church is capable of possessing and exercising the right to freedom of religion in its own capacity as a representative of its members and the entire functioning of churches depends on respect for this right. However, unlike Article 9, Article 8 of the Convention has more an individual than a collective character [].'[12] Subsequently, the Commission declared the complaint inadmissible. In similar fashion, the Court has rejected the capacity of groups to complain about a violation of human rights. Against the intention of the authors of the Convention, it has stressed that only individuals who have been harmed personally and significantly by a specific violation or infringement can bundle their claims. They are approached as a collective, rather than a group.

Consequently, Article 8 ECHR has been so interpreted by the Court that it primarily aims at protecting individual interests by granting individuals a right to complain. However, under certain circumstances, individuals have also been allowed to claim a right if their personal interest is intertwined with the interests of a legal person or of a (minority) group. The most basic example is of course the protection of family life, one of the four terms explicitly mentioned in Article 8 ECHR. If fam-

[9] ECmHR, Tauira a.o./France, appl.no. 28204/95, 04/12/1995.

[10] ECtHR, Asselbourg a.o./Luxembourg, appl.no. 29121/95, 29/061999.

[11] See in further detail: B. van der Sloot (forthcoming a).

[12] ECmHR, Church of Scientology of Paris/France, appl.no. 19509/92, 09/01/1995.

ily life is disturbed, one or all family members may submit a claim under the Convention. Furthermore, the Court has accepted caravans and other mobile homes and temporary shelters under the concept of 'home', which has had important consequences for Gypsies and other nomadic groups,[13] who generally do not possess a fixed shelter or home.[14] For example, the Court has stressed in reference to an applicant that the '[] occupation of her caravan is an integral part of her ethnic identity as a Gypsy, reflecting the long tradition of that minority of following a travelling lifestyle. This is the case even though, under the pressure of development and diverse policies or by their own choice, many Gypsies no longer live a wholly nomadic existence and increasingly settle for long periods in one place in order to facilitate, for example, the education of their children. Measures affecting the applicant's stationing of her caravans therefore have an impact going beyond the right to respect for her home. They also affect her ability to maintain her identity as a Gypsy and to lead her private and family life in accordance with that tradition.'[15]

What is more, states may be under the positive obligation to take active measures to respect and facilitate the development of these minority identities. In *Aksu v. Turkey,* the Court emphasized the '[] emerging international consensus amongst the Contracting States of the Council of Europe, recognising the special needs of minorities and an obligation to protect their security, identity and lifestyle, not only for the purpose of safeguarding the interests of the minorities themselves, but also to preserve a cultural diversity of value to the whole community.'[16] This right to respect for minority life requires states to accept 'that special consideration should be given to their needs and their different lifestyle, both in the relevant regulatory framework and in reaching decisions in particular cases' in order to allow them to fully explore, develop and express their identity, and that governments 'should pursue their efforts to combat negative stereotyping of the Roma', among others, because 'any negative stereotyping of a group, when it reaches a certain level, is capable of impacting on the group's sense of identity and the feelings of self-worth and self-confidence of members of the group. It is in this sense that it can be seen as affecting the private life of members of the group'.[17]

Consequently, individuals are allowed to claim a right to privacy to protect a group interest under the European Convention on Human Rights if their interests collides with or are part of the interest of a (minority) group. Doing so, they can protect the interests of the group as a whole. The same counts for the protection of the interests of legal persons. Strange as it may sound, this seems partly a consequence of the increased focus on the natural person, his private interests and the full development of his personality. In the early jurisprudence of the former Commission and the Court, it was held that a second home, a building site, a professional work-

[13] ECmHR, Lay/UK, appl.no. 13341/87, 14/07/1988.

[14] ECmHR, Smith/UK, appl.no. 14455/88, 04/09/1991. ECmHR, Smith/UK, appl.no. 18401/91, 06/05/1993.

[15] ECtHR, Chapman/UK, appl.no. 27238/95, 18/01/2001, § 73.

[16] ECtHR, Aksu/Turkey, appl.nos. 4149/04 and 41029/04, 27/07/2010, § 49.

[17] ECtHR, Aksu/Turkey, appl.nos. 4149/04 and 41029/04, 15/03/2012, § 58 & 75.

ing place, a temporary shelter or other unconventional houses did not fall under the scope of 'home'. For example, with regard to the search of a person's car, which functioned as his home, the Commission held: '[] la Commission estime que le domicile – "home" – dans le texte anglais de l'article 8 (art. 8)- est une notion précise qui ne pourrait être étendue arbitrairement et que, par conséquent, la fouille de la voiture en stationnement dans les circonstances de la présente affaire, ne saurait être assimilée à une fouille dimiciliaire [sic] qui entre dans le domaine d'application de l'article 8 (art. 8).'[18] However, the Convention is drafted in two official languages, English and French, and the French version of the European Convention does not refer to 'maison', 'chez' or 'residence' but rather to the concept 'domicile'. Domicile has a broader scope than the concept of 'home' and might, for example, be used to refer to professional dwellings.

Building on this line of interpretation, the Court has accepted that individuals who work from home also fall under the scope of Article 8 ECHR, that the interests of one-man firms may be part and parcel of the interest of the natural person and that consequently, entering a business premises (by the police) may affect the right to privacy of a natural person. Likewise, the secrecy of communication, as guaranteed by Article 8 ECHR, is held by the Court to extent to professional communications.[19] The Court has held, among other cases, in its *Chappell* and *Niemitz* judgments, that there is no reason of principle why the notion of private life should be taken to exclude activities of a professional or business nature. 'This view is supported by the fact that [] it is not always possible to distinguish clearly which of an individual's activities form part of his professional or business life and which do not. Thus, especially in the case of a person exercising a liberal profession, his work in that context may form part and parcel of his life to such a degree that it becomes impossible to know in what capacity he is acting at a given moment of time.'[20] Thus, the professional life of an applicant and the interests of a legal person (for example his one-man business) may also affect a person's private life and consequently form a part of his personal interest. Consequently, individuals can also submit a complaint to protect the interests of legal persons, if their personal interests are part of or collide with those interests. Doing so, they can protect the interests of the legal person as such.

Interest				
Right	Individual	Legal person	Group	General
Individual	x	x	x	
Legal person				
Group				
State	x	x	x	x

[18] ECmHR, x./Belgium, appl.no. 5488/72, 30/05/1974.

[19] ECtHR, Chappell/UK, appl.no. 10461/83, 30/03/1989. ECtHR, C./Belgium, appl.no. 21794/93, 07/08/1996, § 25.

[20] ECtHR, Niemietz/Germany, appl.no. 13710/88, 16/12/1992, § 25.

11.4 General Interests

As has been noted in Sect. 11.2, the origins of the Convention lie in the protection of general interests, related to the abuse of power by states; consequently, individuals, legal persons, groups and state could initiate a complaint procedure regarding a violation of Article 8 ECHR. As noted in Sect. 11.3, the Court has reinterpreted the right to privacy and has stressed that this doctrine, in principle, only protects individual interests and only grants individuals a right to complain. Inter alia, it rejects *in abstracto* claims, hypothetical harm and class actions, which do not aim to protect the specific interests of claimants, but are related to the general interest, for example regarding the abuse of legislative or administrative power as such. However, in exceptional cases, the Court is prepared to relax its focus on individual interests and individual rights and accepts claims protecting the general interest, particularly if they concern mass (secret) surveillance by states.

The first case in which it did so was *Klass and others v. Germany*, in which the applicants challenged the German legislation in that it permitted covert surveillance measures without obliging the authorities in every case to notify the persons concerned after the event, and in that it excluded any remedy before the courts against the ordering and execution of such measures. This led, according to them, to a situation of potentially unchecked and uncontrolled surveillance, as those affected by the measures were kept unaware and would thus not challenge them in a legal procedure. In essence, the case revolved around hypothetical harm, as the applicants claimed that they could have been the victims of surveillance activities conducted by the German government, but they were unsure as the secret services remained silent on this point. The Commission, deciding on the admissibility of the case, referred to Article 25 ECHR, the current Article 34 ECHR. It argued that under this provision 'only the victim of an alleged violation may bring an application. The applicants, however, state that they may be or may have been subject to secret surveillance, for example, in course of legal representation of clients who were themselves subject to surveillance, and that persons having been the subject of secret surveillance are not always subsequently informed of the measures taken against them. In view of this particularity of the case the applicants have to be considered as victims for purposes of Art. 25.'[21]

Before the Court, who dealt with the case in substance, the Delegates of the Commission considered that the government was requiring a too rigid a standard for the notion of 'victim'. They submitted that, in order to be able to claim to be the victim of an interference with the exercise of this right, 'it should suffice that a person is in a situation where there is a reasonable risk of his being subjected to secret surveillance.'[22] The Court, however, took it one step further and held that 'an individual may, under certain conditions, claim to be the victim of a violation occasioned by the mere existence of secret measures or of legislation permitting secret

[21] ECmHR, Klass a.o./Germany, appl.no. 5029/71, 18/12/1974.

[22] ECtHR, Klass a.o./Germany, appl.no. 5029/71, 06/09/1978, § 31.

measures, without having to allege that such measures were in fact applied to him.'[23] In this case, the Court thus accepted an *in abstracto* claim, instead of hypothetical damage, as the 'mere existence' of a law may lead to an interference with Article 8 ECHR. This contrasts with the test proposed by the Delegates, namely whether there is a 'reasonable likelihood' that the applicants were affected by the measures complained of. In the latter test, the requirement of personal harm remains, though it is not made dependent on actual and concrete proof, but on a reasonable suspicion; in the abstract test, the requirement of personal harm is abandoned, as the laws and policies are assessed as such.

The abstract test has since then been accepted in a handful of cases.[24] Importantly, in *in abstracto* claims, the Court lets go of the requirement of individual harm of the claimant and is willing to allow any party to submit a complaint on behalf of society. This means that not only natural persons (note: the victim requirement is abandoned here) are allowed to rely on the right to privacy, but legal persons as well. In *Mersch and other v. Luxembourg*, for example, the Commission carefully distinguished between the two tests, applying the abstract and the hypothetical test to two different types of complaints. The case was declared incompatible with the provisions of the Convention, in so far as it regarded a violation of the Convention's provisions on account of measures taken under a legal instrument, as the claimants had not been subjected to surveillance measures. Likewise, it stressed that legal persons could not complain about such matters as they could not be subjected to monitoring or surveillance ordered in the course of criminal proceedings because legal persons have no criminal responsibility. However, part of the claim regarded laws allowing for surveillance not confined to persons who may be suspected of committing the criminal offences referred to therein. With regard to this abstract claim, the Commission received both natural and legal persons and declared the case admissible.[25] This was confirmed in later jurisprudence.[26] Consequently, both individuals and legal persons can, under certain circumstances, submit a claim under Article 8 ECHR in order to protect a general, societal interest. Groups, however, have not been able to do so so far.[27]

[23] ECtHR, Klass a.o./Germany, appl.no. 5029/71, 06/09/1978, § 34.

[24] See in further detail: van der Sloot (2016).

[25] ECmHR, Mersch a.o./Luxembourg, appl.nos. 10439/83, 10440/83, 10441/83, 10452/83, 10512/83 and 10513/83, 10/05/1985.

[26] ECtHR, Liberty a.o./UK, application no. 58243/00, 01/07/2008, § 56–57. ECtHR, Association for European Integration and Human Rights and Ekimdzhiev/Bulgaria, appl.no. 62540/00, 08/06/2007, § 59. ECtHR, Iordachi a.o./Moldova, appl.no. 25198/02, 10/02/2009, § 33–34. See also: ECtHR, Telegraaf Media Nederland Landelijke Media B.V. a.o./Netherlands, appl.no. 39315/06, 22/11/2012.

[27] Of course, one could argue that because the victim requirement is abandoned, the natural persons complaining under the ECHR are a group, because they are not separable on the ground of their individual, personal interests. Rather, they have a shared interest in not having a shared interest, namely being subjected to a certain law or policy. This would mean that the whole population of a country would be a group. Whether this would count as 'group privacy' is debatable – this will be further discussed in the analysis of this chapter.

Interest				
Right	Individual	Legal person	Group	General
Individual	x	x	x	x
Legal person				x
Group				–
State	x	x	x	x

11.5 Legal Persons' Right to Invoke the Right to Privacy

Section 11.2 discussed under which circumstances states could pursue the claims of legal persons, protecting the interests of those legal persons. Section 11.3 described in which cases individuals are allowed to submit a complaint in order to protect the interests of a legal person. Section 11.4 then analysed under which conditions legal persons are allowed to submit a claim on behalf of society, to protect a general interest. Section 11.3, however, also noted that in principle, legal persons are not allowed to invoke Article 8 ECHR, in contrast to invoking other rights under the European Convention on Human Rights, as the Court has stressed that the right to privacy, in contrast to those other rights, only protects individual, and not general interests. As legal persons have no human dignity, personal autonomy or individual freedom, they are in principle not allowed to invoke Article 8 ECHR. Still, however, Sect. 11.3 also described the Court's tendency to stretch the boundaries of the Convention and its strictly limited focus on individual interest and individual rights. Building on its interpretation that 'domicile', the French term for 'home', may also include business premises, the Court has accepted that individuals may also protect the interests of a legal person if they are intertwined with the interests of legal persons.[28]

In a case from 2002, *Stes Colas Est and others v. France*, in which a business complained about the searches and seizures of the government on their business premises, the Court was prepared to go even further and changed its position on the admissibility of complaints by legal persons 'In *Chappell v. the United Kingdom*, the Court considered that a search conducted at a private individual's home which was also the registered office of a company run by him had amounted to interference with his right to respect for his home within the meaning of Article 8 of the Convention. The Court reiterates that the Convention is a living instrument which must be interpreted in the light of present-day conditions. [...] Building on its dynamic interpretation of the Convention, the Court considers that the time has come to hold that in certain circumstances the rights guaranteed by Article 8 of the Convention may be construed as including the right to respect for a company's registered office, branches or other business premises.'[29] Although privacy claims by

[28] See more elaborate on this topic: van der Sloot (2015).

[29] ECtHR, Stes Colas Est a.o./France, appl.no. 37971/97, 16/04/2002, § 40–41.

legal persons have subsequently only been accepted in a very limited number of cases, and are mostly rejected,[30] in later jurisprudence, the ECtHR has confirmed its position that under certain circumstances, businesses may successfully invoke a right to privacy to protect their own interests.[31]

However, it is less eager to allow legal persons to rely on Article 8 ECHR in order to protect the privacy interests of individuals or groups. For example, the former Commission was faced with a complaint by a church about the refusal of access to files which it assumed were held on it by the Criminal Intelligence Department of the French Ministry of the Interior and which it believed were likely to contain wrongful data on it, its officials and its members. 'The Commission does not find it necessary in the present case to examine exhaustively to what extent a legal person may invoke the right to respect for private life within the meaning of Article 8 of the Convention. It observes that the applicant association seeks access to files containing data of its members, i.e. of private persons. It recalls that the French authorities do not deny private persons access to data concerning them. In these circumstances, even assuming that Article 8 of the Convention applies, the Commission does not consider that to deny the applicant association access to those files constitutes a lack of respect for the applicant association's own private life. The Commission accordingly finds no appearance of a violation of Article 8 in this respect.'[32] Consequently, if natural persons are capable of invoking Article 8 ECHR to protect their own interests, legal persons cannot do so on their behalf.

Similarly, the Court in principle rejects claims by legal persons to protect the interests of groups. In contrast, it has allowed individuals to submit class actions, to protect the interests of groups.[33] For example, in *Marckx v. Belgium*, the inheritance laws complained of had not yet been applied to the applicants and presumably would not be applied for a certain period of time, but the Court argued nonetheless that they had a legitimate interest in challenging a legal position, that of an unmarried mother and of children born out of wedlock, which affected them – according to the Court – personally.[34] In *Dudgeon v. the United Kingdom*, the case regarded a claim by an applicant about the regulation of homosexual conduct. The Court held that the applicant could be received even without the law being applied to him and without there being any reason to believe that it might be, as 'the very existence of this legislation continuously and directly affects his private life: either he respects the law and refrains from engaging – even in private with consenting male partners – in prohibited sexual acts to which he is disposed by reason of his homosexual tendencies, or he commits such acts and thereby becomes liable to criminal

[30] ECtHR, Vallianatos a.o./Greece, appl.nos. 29381/09 and 32684/09, 07/11/2013. ECtHR, Winterstein a.o./France, appl.no. 27013/07, 17/10/2013. ECtHR, Avilkina a.o./Russia, appl.no. 1585/09, 06/06/2013.

[31] ECtHR, Wieser and Bicos Beteiligungen GMBH/Austria, appl.no. 74336/01, 16/10/2007. ECtHR, Saint-Paul Luxembourg S.A./Luxembourg, appl.no. 26419/10, 18/04/2013.

[32] ECmHR, Church of Scientology of Paris/France, appl.no. 19509/92, 09/011995.

[33] ECmHR, Brüggemann and Scheuten/Germany, appl.no. 6959/75, 19/05/1976.

[34] ECtHR, Marckx/Belgium, appl.no. 6833/74, 13/06/1979, § 27.

prosecution.'[35] However, even in these types of cases, where the victim requirement is relaxed, legal persons are in principle not allowed to rely on Article 8 ECHR.[36]

Still, even on this point, the Court is occasionally willing to relax its position, especially in more recent case law. For example, in a case from 2013, three companies complained that their right to respect for privacy, home and correspondence under Article 8 of the Convention had been infringed as a result of the Supreme Court's judgement upholding the Directorate of Taxation's decision of 1 June 2004. Among others, the Government argued that, whilst the applicant companies had maintained that the backup copy of the server had contained e-mails to and from different people working for the applicant companies and that an inspection of the tape would interfere with their "legitimate right for privacy at work", no one working for them had complained before the Court. 'The matters which the applicant companies were pursuing under the Convention concerned natural persons working for them, not the companies themselves. Thus the applicant companies could not be regarded as "victims" within the meaning of Article 34. The Government invited the Court to declare this part of the application inadmissible as being incompatible ratione personae.' The Court, however, noted 'that the applicant companies' interest in protecting the privacy of their employees and other persons working for them did not constitute a separate complaint but only an aspect of their wider complaint under Article 8 of the Convention. The fact that no such individual person was a party to the domestic proceedings nor brought an application under the Convention should not prevent the Court from taking into account such interests in its wider assessment of the merits of the application.'[37]

Similarly, the Court is prepared to receive legal persons who submit a complaint on behalf of a certain group, inter alia, trade unions or professional associations representing the interests of those exercising a certain profession. For example, the general national association for journalists could successfully rely on Article 8 ECHR to protect the interests of specific journalists and of the journalistic group as a whole, in a case which regarded laws that granted the administrative power very broad and wide powers to search premises and wire-tap telecommunication.[38] The same counts, among others, when lawyers are subjected to such strict measures. Legal persons are then allowed to rely on Article 8 ECHR, even when no specific individuals have been affected by a certain law or policy to protect the interests of the group.[39] It is unsurprising that the Court allows such complaints in relation to these types of professions. Lawyers, journalists and doctors cannot exercise their profession without a certain amount of secrecy being guaranteed – the privacy between clients and lawyers, the secrecy of sources for journalists and the confiden-

[35] ECtHR, Dudgeon/UK, appl.no. 7525/76, 22/10/1981, § 41.

[36] ECmHR, Norris, National Gay Federation/Ireland, appl.no. 10581/83, 16/05/1985. ECtHR, Vallianatos a.o./Greece, appl.nos. 29381/09 and 32684/09, 07/11/2013.

[37] ECtHR, Bernh Larsen Holding AS a.o./Norway, appl.no. 24117/08, 14/03/2013.

[38] ECtHR, Ernst a.o./Belgium, appl.no. 33400/96, 15/07/2003.

[39] ECtHR, André a.o./France, appl.no. 18603/03, 24/07/2008.

tiality between patients and doctors are all a *conditio sine qua non* for exercising these professions.

Interest / Right	Individual	Legal person	Group	General
Individual	x	x	x	x
Legal person	x	x	x	x
Group				–
State	x	x	x	x

11.6 A Group's Right to Invoke the Right to Privacy

The final question is whether groups are allowed to rely on Article 8 ECHR to protect either their own interests or that of others, such as natural persons or legal persons. This is a difficult question to answer, because the legal system in general tends to stimulate groups to obtain a legal status and a form of hierarchy. This ensures that the government, the judge or any other organisation knows who is the representative of a certain group or minority. Suppose that a specific group or community was stigmatised by a law or policy and it wanted to challenge its position before the court: should the whole community submit a complaint, should each and every member join as an individual claimant to this complaint, should it create a legal organisation in order to represent it? Rather quickly, the tendency is to choose either a form in which individual complaints are bundled and aggregated, in which one individual is said to represent the whole group or in which a legal organisation has the task of legal representation. Thus, the tendency is to move to collective or corporate rights, but the question is whether there is also be a right of groups to invoke a right to privacy itself.

The ECtHR in principle rejects cases in which individuals bundle their complaints or submit a complaint on behalf of a group, without the applicants being harmed themselves. As a typical example, one might take the case of *Stankov, Trayanov, Stoychev, United Macedonian Organisation "Ilinden", Mechkarov and others v. Bulgaria*, in which the applicant association and the individual complainants argued that Bulgaria did not recognise the existence of a Macedonian minority and of the Macedonian nation. Inter alia, in the forms for the 1992 census of the population, in the space provided for declaring one's ethnic origin, there was no mention of a Macedonian ethnic origin. Also, the Macedonian language could not be used in the relations with the administration, the Ministry of Education had refused to introduce the study of the Macedonian language and history in State schools and the applicants complained about the alleged general campaign in the media against them. The Commission recalled right away that it did not allow 'any form of "actio popularis" but requires, when the right to individual petition under Article 25 is being exercised, that the applicant can claim to be the victim of a violation of the Convention. The Commission, therefore, can consider each of the applicants' complaints only insofar as those who have raised the respective complaint can claim to be, personally, victims of a particular violation of their rights under the Convention.' The Commission went on to note 'that the only particular fact submitted by the applicants is the lack of a special mention of a Macedonian ethnic origin in the forms for the 1992 census of the population. None of the applicants has shown that the alleged non-recognition of a Macedonian minority engenders for him or her such direct practical consequences as to amount to an interference with the right, for example, to respect for a person's private life under Article 8 of the Convention, or with the other rights guaranteed by the Convention. It follows that the above complaint has to be rejected under Article 27 para. 2 of the Convention.'[40]

On the other hand, the Court does allow people to bundle their claims, even if they form very large groups. For example, the very first cases that were declared admissible under Article 8 of the European Convention on Human Rights regarded complaints by groups.[41] The two cases were *Habitants D´Alsemberg, de Beersel, de Kraainem, d´Anvers et Evirons, de Grand et Environs v. Belgium* and *Habitants de la Région des Fourons v. Belgium.*[42] The titles of these cases already suggests that the applicants are seen as a group, rather than a collective of individuals who so happen to share a similar interest. The latter case regarded a complaint by an association against the Belgian Government regarding the violation of Articles 8 and 14, on behalf of 165 fathers of 311 children. These parents lived in six communes, that made up the area known as the "Voerstreek", located in the north-east of Liège. The case regarded the school system of Belgium, in which the schools in Wallonia were French speaking and those in Flanders were Dutch speaking. This forced parents who wanted their children to be educated in another language than was common in that region either to accept that their children would be raised in another language, to move to another part of the country or to send their children away from home. This, they argued, constituted a violation of their right to respect for family life. The case submitted by the association was declared admissible. The other case regarded a similar complaint and it led to the first case before the European Court of Human Rights regarding a possible violation of Article 8 ECHR, called the *Case "Relating to certain aspects of the law on the use of languages in education in Belgium" v. Belgium*, and although the group of claimants was received in its claim, the Court found no violation of the right to privacy.[43] Reference could also be made to the case

[40] ECmHR, Stankov a.o./Bulgaria, appl.nos. 29221/95, 29222/95, 29223/95, 29225/95 and 29226/95, 21/10/1996.

[41] See also: ECmHR, Un Groupe D'Habitants De Leeuw-st-Pierre/Belgium, appl.no. 2333/64, 16/12/1968. ECmHR, Confederation des Syndicats Medicaux Francais et Federationale Des Infirmiers/France, appl.no. 10983/84, 12/05/1986.

[42] ECmHR, Habitants de la Région des Fournons/Belgium, appl.no. 2209/64, 15/12/1964.

[43] ECtHR, Relating to certain aspects of the law on the use of languages in education in Belgium/ Belgium, appl.nos. 1474/62, 1677/62, 1691/62, 1769/63, 1994/63 and 2126/64, 23/07/1968.

of *Moldovan and others v. Romania (no. 2),* in which seven individuals complained about the discrimination they were exposed to because of their Roma origin.[44]

This case, like a few cases referred to in Sect. 11.3, hints towards an important point. Although these cases are brought forward by a specific individual or several individuals, who are part of a certain minority group, the effects of the judgments are far broader. This already is the case when a matter regards a negative obligation of the state, such as stopping a certain policy or revoking a certain law, but is even more prominent in cases in which the Court decides on a positive obligation of the state. In the cases discussed earlier, such as *Aksu,* the Court accepts a positive obligation of the state to relief the disadvantaged position of a group, among others, by actively protecting the reputation, position and property of minority groups such as Gypsies, Kurds, or Jews. Thus, although the claim is brought forward by individuals, a judgment by the Court may have significance for a group or minority as a whole. There are a few other interesting cases to note in this respect. These regard matters in which a group of persons has a similar interest in respecting privacy.

One might for instance point to data collection[45] and wire-tapping, which may affect a group of people.[46] For example, the case of *Petri Sallinen and others v. Finland*, which regarded a claim by 18 Finnish nationals, who complained about the search and seizure of privileged material in the first applicant's law firm. The first applicant was the lawyer, the other claimants were his clients. The clients claimed that the search and seizure of privileged material interfered with their rights under Article 8 ECHR. The Government contested their view, arguing that even though the correspondence with a lawyer falls under the protection of Article 8, there had not been any interference with their rights, only with the rights of the first applicant. The Court disagreed, finding 'that the search by the police of the residential premises and the business premises of the first applicant, and the seizure of hard disks there, amounted to an interference with the right to respect for the first applicant's "home" and "correspondence", as those terms have been interpreted in the Court's case-law. It follows that the search and seizure also amounted to an interference with the right to respect for the client applicants' "correspondence".'[47]

Thus, a particular violation might affect a group interest in privacy. Of course one could argue that it is still individual claimants here that invoke the right to privacy, who all rely on the right to privacy for the protection of their own interests, the interests being different for the lawyer and his clients. Although this is true, it is also important to stress that this group is founded by and depended on the notion of secrecy – without their secrecy being respected, this group would not have formed or would have taken a different form. The same counts, as has been stressed earlier, for the relationship between patients and doctors, between journalists and their sources, etc. If it is true that these relationships are built and dependent on the

[44] ECtHR, Moldovan a.o./Romania (no. 2), appl.nos. 41138/98 and 64320/01, 12/07/2005.

[45] ECtHR, Stedt-Wiberg a.o./Sweden, appl.no. 62332/00, 06/06/2006.

[46] ECtHR, K. H. a.o./Slovakia, appl.no. 32881/04, 28/04/2009.

[47] ECtHR, Petri Sallinen a.o./Finland, appl.no. 50882/99, 27/09/2005, § 71.

protection of privacy, the violation of privacy is something more than an aggregated interests of several individuals. It is a constitutive element of the group as such.

Finally, there are certain cases regarding environmental protection which could be viewed from the perspective of group rights. Environmental pollution has been granted protection by the Court under the scope of Article 8 ECHR when it affects someone's home, family life and especially private life. This might be the cases, for example, when night flights disturb one's sleep, when fumes diminish the quality of life and when dangerous fumes and smog might affect the health of persons. The point is that many of these issues affect a larger group of people, sometimes whole towns. This, perhaps, does not yet make them a group, but it seems that they have something more in common than a shared individual interest. This point is strengthened by the fact that in these types of cases, the Court is willing to relax its position on the requirement of individual harm and individual interests.

The Court has done something peculiar in these types of cases, namely it has adopted as essential notion whether the 'quality of life' of the applicant has been harmed. The issue with many environmental cases, particularly, is that the notion of harm is so problematic. What harm does noise pollution do to the individual's private or family life? How can one substantiate, for example, that medical problems have arisen from smog or air pollution? What harm does the greenhouse effect do to the individual complainant? Even if personal harm can be demonstrated, the causal connection between environmental pollution and individual harm is often very difficult to demonstrate. This is exactly why the Court has introduced the notion of 'quality of life', as whether the 'quality of life' is diminished can in principle only be determined by the subject itself. Thus, the notion of harm becomes a subjective, rather than an objective matter. 'The "quality of life" is a very subjective characteristic which hardly lends itself to a precise definition.'[48] It is this term that allowed the Court to declare the case admissible. Consequently, the criterion used by the Court in these types of cases is not whether the individual complainant is harmed as such, but whether he beliefs he has been harmed. This has allowed larger groups of people to submit a complaint, sometimes more than 300, and has allowed the Court to accept the applicants without having to assess whether and if so, to what extent each and every individual complainant has suffered from the privacy violation complained of.

Finally, reference should be made again to the fact that positive obligations play an important role. For example, *Guerra and others v. Italy* revolved around the claim of a group of inhabitants of a town located nearby a polluting factory. The Court reiterated 'that severe environmental pollution may affect individuals' well-being and prevent them from enjoying their homes in such a way as to affect their private and family life adversely. In this case the applicants waited, right up until the production of fertilisers ceased in 1994, for essential information that would have enabled them to assess the risks they and their families might run if they continued to live at Manfredonia, a town particularly exposed to danger in the event of an accident at the factory. The Court holds, therefore, that the respondent State did not

[48] ECtHR, Ledyayeva a.o./Russia, appl.no. 53157/99, 53247/99, 53695/00 and 56850/00, 26/10/2006, § 90.

fulfil its obligation to secure the applicants' right to respect for their private and family life, in breach of Article 8 of the Convention. There has consequently been a violation of that provision.'[49] Consequently, the state may have a duty to inform a group of people that might be affected by certain environmental pollution. Likewise, the Court has held in a number of cases that the group of people likely to be affected by certain changes in the environment (the creation of a factory, the building of an airport, the construction of a road, etc.) must be allowed to take part in the decision-making process concerning that environmental change.[50]

In conclusion, it seems that the Court is not yet ready to accept groups as such as claimants of the right to privacy, either in order to protect their own interests or to protect those of others, such as individuals or legal persons. Still, there are several points which could be used by the Court at a later stage to change its stance on this point, as it did when it changed its position on the admissibility of legal persons. Although the Court has not yet been prepared to make the final step with regard to group privacy, it has accepted that large groups of people and indeed sometimes whole towns have a right to complain, that privacy protects the fundaments of professional groups of, inter alia, lawyers and clients, doctors and patients and journalists and their sources, and that positive obligations for states may entail the active protection of certain minority groups in society.

Interest Right	Individual	Legal person	Group	General
Right				
Individual	x	x	x	x
Legal person	x	x	x	x
Group	–	–	–	–
State	x	x	x	x

11.7 Analysis

This chapter has analysed whether groups are allowed to invoke a right to privacy under the ECHR to protect their own interests. It has also described in how far individuals, legal persons and states can rely on Article 8 of the Convention to protect the privacy interests of groups. It appears that although others are mostly allowed to invoke the right to privacy to protect the interests of groups, groups themselves are not allowed to do so. It also appeared that although there are a number of exceptions, the dominant focus of the European Court of Human Rights, as described in

[49] ECtHR, Guerra a.o./Italy, appl.no. 14967/89, 19/02/1998.

[50] ECtHR, Hatton a.o./UK, appl.no. 36022/97, 08/07/2003. ECtHR, Taskin and others v. Turkey, application no. 46117/99, 10 November 2004. ECtHR, Ockan a.o./Turkey, appl.no. 46771/99, 28/03/2006. ECtHR, Di Sarno a.o./Italy, appl.no. 30765/08, 10/01/2012. ECtHR, Kolyadenko a.o./Russia, appl.nos. 17423/05, 20534/05, 20678/05, 23263/05, 24283/05 and 35673/05, 28/02/2012.

Sect. 11.3, is on natural persons who have an individual right to protect their personal interests related to human dignity, individual autonomy and personal freedom. This paradigm has functioned relatively well for decades as most privacy violations were targeted at specific individuals. However, the current technological reality, often referred to as big data, means that threats to privacy increasingly do not take place on an individual level, but on a general or group level. Should then groups be allowed to invoke a right to privacy to protect their group interest?

There are several reasons to answer this question affirmatively. There seems an increasingly big chasm between the technological developments and the juridical paradigm.[51] The right to privacy is perhaps the concept where this divide is most visible. In short, the current privacy paradigm grants a natural person a subjective right to claim his right to privacy, this right protects his legitimate interests to human dignity, personal autonomy or individual freedom and in concrete cases, these private interests are balanced against the common interest involved with a privacy violation, such as national security. It seems that this focus on individual rights and individual interests of natural persons no longer holds in an age of big data, or whatever term is used to capture the societal tendency to collect, store, analyse and use massive amounts of data for all kinds of purposes and policies. Of course, subjective rights and individual interests will always remain of (the greatest) importance – if nude pictures leak, if a person is spied upon for years, if his house is entered by his neighbours without asking – a person should always have a subjective right to protect his individual interests. But the current legal paradigm is relatively fit for addressing these types of problems, although it may require some dusting here and there. What is essential to these new technological developments, however, is that they do not revolve around the individual and his specific interests.

Big data may be defined as gathering massive amounts of data without a pre-established goal or purpose, about an undefined number of people. These data are then processed on a group or aggregated level through the use of statistical correlations. The essence is thus that the individual element is mostly lost. Data are not gathered about a specific person or group (for example those suspected of having committed a particular crime), rather, they are gathered about an undefined number of people during an undefined period of time without a pre-established reason. The potential value of the gathered data becomes clear only after they are subjected to analysis by computer algorithms, not beforehand. These data, even if they are originally linked to specific persons, are subsequently mainly processed on an aggregated level by finding statistical correlations. It may appear that the data string – Muslim + vacation to Yemen + visit to website X – leads to an increased risk of a person being a terrorist. The data are not based on personal data of specific individuals, but processed on an aggregated level and the profiles revolve around groups. (Note: if a specific individual is discriminated upon on the basis of a general profile, this has an impact on his individual interests and subjective rights – but the problem of the creation of the profile itself and the fact that policies are based on such profiles remains unaddressed. This becomes even more urgent when these

[51] See more elaborately: van der Sloot forthcoming b).

profiles are not based on sensitive data nor lead to severe restrictions, but are based on general data, zip codes for example, and are used to develop social and economic policies. This might be problematic, because people are judged and treated according to pre-established profiles and pre-established character traits, but the harm to the specific individual is difficult to demonstrate).

Given this constellation of facts, it becomes more and more difficult for an individual to point out his specific personal interest and personal harm in the technological reality. It should be acknowledged that in the field of privacy, the notion of harm has always been problematic as it is often difficult to substantiate the harm a particular violation has done, e.g. what harm follows from entering a home or eavesdropping on a telephone conversation as such when neither objects are stolen nor private information disclosed to third parties? Even so, the more traditional privacy violations (house searches, telephone taps, etc.) are clearly demarcated in time, place and person and the effects are therefore relatively easy to define. In the current technological environment, however, the individual is often simply unaware that his personal data are gathered by either his fellow citizens (e.g. through the use of their smartphones), by companies (e.g. by tracking cookies) or by governments (e.g. through covert surveillance). Obviously, people unaware of the fact that their data are gathered will not invoke their right to privacy in court.

But even if a person would be aware of these data collections, given the fact that data gathering and processing is currently so widespread and omnipresent, and will become even more so in the future, it will quite likely be impossible for him to keep track of every data processing which includes (or might include) his data, to assess whether the data controller abides by the legal standards applicable, and if not, to file a legal complaint. And if an individual does go to court to defend his rights, he has to demonstrate a personal interest, i.e. personal harm, which is a particularly problematic notion in big data processes, e.g. what concrete harm has the data gathering by the NSA done to an ordinary American or European citizen? In these types of cases, the problem is not that this or that specific person has been affected, or that his specific interests have been harmed, but that large groups are or society as a whole is affected and that group or societal interests are undermined. For example, the problem with the NSA affair or hanging CCTV cameras on the corner of every street is not that specific individuals are affected, rather such initiatives pose a structural question regarding how power is used (or perhaps better, abused). These large data-gathering systems and mass surveillance activities by states undermine the trust people have in governmental institutions and perhaps more importantly, undermine the minimum conditions for the legitimate use of power. Using the state's power to surveil so many people, at so many places over so many years without a clear and concrete reason, simply verges on the abuse of power.

As has been said, the right to privacy is perhaps the right where the tension between the technological developments and the current legal paradigm is most visible. But other rights face a similar problem, perhaps most prominently the right to data protection. This right is also (increasingly) based on the idea of individual rights to control data (through doctrines such as informed consent, the right to be forgotten and the right to data portability) and to seek legal remedy by invoking subjective rights. And similar to privacy, data protection aims at protecting individual interests – the scope of data protection instruments is determined by the term 'personal data' which is defined as any data that can be used to identify a natural person. But here too, the problem is that the data that are gathered often do not directly identify a person, but are gathered, assessed and used on a general, aggregated or group level. For example, they may be used to adopt policies on the basis of zip codes, of income levels, or of any other general criterion. These data do thus not directly identify a person, and consequently fall outside the scope of the data protection regulations, although they may affect him as being part of a specific group. (Please note: of course one could focus on the initial moment when personal data are gathered and not yet aggregated, but this may only concern the split second which it takes to aggregate data. The same counts for the moment at which group profiles are applied and used to affect a specific person. This only concerns the very end of the data process. By focusing on the individual, his interests and rights, one loses sight of the larger part of the data processing scheme and the general issues concerned with that).

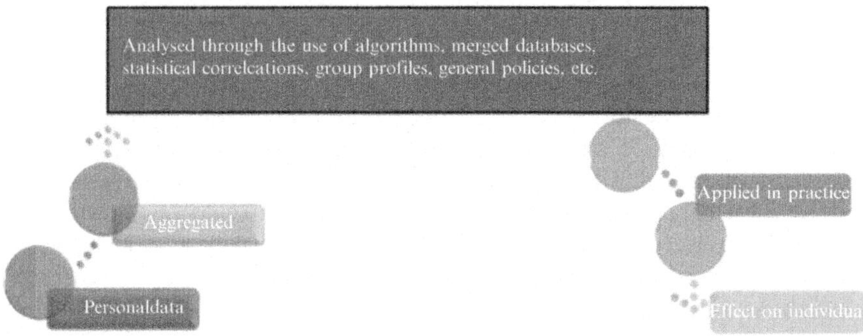

This trend has an effect on other human rights too, for example the right to be free from discrimination. Of course the policies described may have a direct effect on the level of the individual. If a person is denied a loan because a bank has calculated that in neighbourhood X, there is an increased risk of people not paying their dues, this has a clear effect on him (this may be even worse, if done through the technique known as red-lining, by which banks may effectively discriminate on the basis of race, because it so happens that in certain neighbourhoods, there is a dominant population of African origin, for example). The same counts for a health insurer who demands a higher monthly payment because a person is part of a group (for example lower educated, male, living in a poor neighbourhood) which is more likely to have an unhealthy lifestyle. This also applies to states, secret service agencies or the police who may decide to follow a person on the basis of the fact that he is a Muslim, visited Yemen recently and goes to a mosque now and then. This all has a clear effect on the individual – as has been said, this was, is and will remain of the greatest importance. But the current legal paradigm is relatively well suited to address these kinds of problems, because it grants people an individual right not to be discriminated against and to go to court if their specific interests have been harmed. The more general problem, however, is difficult to approach from the existing paradigm (a peculiarity of the discrimination clause under the European Convention on Human Rights, Article 14, is that the Court has decided that this provision may only be invoked in combination with one of the subjective rights granted under the ECHR, such as the right to privacy, the freedom of expression and the freedom of religion). The fear is that groups are stigmatized and that the characteristics of a person may be fixed because he will be profiled by states, businesses and citizens.[52] This also feeds the thought that this will create an increased division between the rich and the poor, as these types of systems tend to give benefits to those in 'good' groups, while avoiding or limiting the claims of 'bad' groups. This can of course be brought down to the level of the individual, but in reality, the problem is societal. It concerns the general issue of a segmented society, which may not only be bad for individual persons, but for society as a whole. (This also brings up the following problem: is this still a legal issue, or is it in reality a political/ethical dilemma?)

To provide a final example, freedom rights, such as the freedom of expression, the freedom of religion, the freedom of assembly and the freedom of movement, are formulated at the individual level, while the restriction on the freedom and autonomy of citizens is increasingly taking place on a group and general level. This is for example the case with city planning based on Big Data processes (smart cities), which affects the environment in which people live without them being aware that certain choices are made to affect their behaviour tacitly. This relates to the debate about nudging (Thaler and Sunstein 2009)– of course the state has a legitimate reason to persuade people to live healthy, among other concerns, and it has always done so, inter alia through taxes on cigarettes, alcohol, etc. But the trend is to move increasingly towards a situation in which it is not so clear for citizens/consumers which of their choices actually are affected and in what way. And again, even if they would be aware, to focus on this observation would be missing the point, not only because these nudges will presumably be so systemic to decision-making in the future that it becomes impossible for individuals to take those nudges into account, but also because it is not that specific choices are affected, but that the whole environment in which people grow up and live is shaped and designed on the basis of large data processes, group profiles, statistical correlations, etc. and the policies that are based on them. For example, there are now plans for designing cities in such a way that if it is known right now that there is an increased risk of obesity in certain neighbourhoods in 20 years' time, squares are designed in a way to motivate people to walk rather than to take the car, elevators in buildings are moved to the back, while stairs are right at the entrance, etc. Here again, it is very difficult to point to a specific individual interest being at stake, a specific form of autonomy being undermined – what harm does it do when a person is motivated to take the stairs rather than the elevator? But it does point to a more general concern and interest, there is a trend to move towards a world in which citizens are constantly and systemically nudged, not only by states and governmental institutions, but also by businesses and perhaps in the future even by fellow citizens.

---

[52] See also: Pariser (2011).

Consequently, there is a constant tension between the level on which the violation takes places and the level at which the legal remedies are provided. This seems to argue in favour of transposing the current legal paradigm to a group level. There are, however, also several counterarguments. A group differs from a legal association on two points (Bygrave 2002). First, who belongs to it is often not clearly demarcated. Second, there is not one univocal answer to the question of what actually is the interest or desire of the group. It is often unclear who decides on such issues – the legal paradigm, in this sense, forces groups to become legally organized, because in such cases it is less difficult to determine who belongs to the group (namely members) and who represents them (legal organizations tend to have a board, a leader or a chairman of some kind). There is one further aspect to keep in mind when discussing group rights in terms of profiling and big data: these groups are not stable, but fluid, and not unique or sparse, but omnipresent and widespread. Group profiles may be created in a split second, they may be used by all kinds of organizations and institutions and they may change, by altering the determinants and criteria according to new insights or needs, so that who is part of a group profile and who is not may change every day, or even more often. These are a few of the most serious problems with granting a group right to privacy.

Perhaps, inspiration could be drawn from a number of other already existing legal doctrines, such as minority rights, relational privacy, future generation rights and the concept of wrongful life. However, there comparison between group privacy and these concepts seems wanting on a number of points. With respect to minority rights as an inspiration for group privacy, three points must be made (Lerner 1991; Raikka and Aikk 1996). First, minority rights are mostly connected to mostly objectively verifiable characteristics. Article 27 of the International Covenant on Civil and Political Rights, for example, provides: 'In those States in which ethnic, religious or linguistic minorities exist, persons belonging to such minorities shall not be denied the right, in community with the other members of their group, to enjoy their own culture, to profess and practice their own religion, or to use their own language.' With group profiling, the profile may be based on such characteristics, but are often also based on more fluid and contingent factors, such as postal code, health status, interest in sport, etc. Second, minority rights have both a negative and a positive aspect – on the negative aspect, it entails that minorities should not be stigmatized, discriminated against, etc. This, however, is not as such an independent doctrine, it is merely an application of the general prohibition on discrimination. It is this negative aspect which may be applicable to group profiling, i.e. not being profiled or discriminated against on the basis of certain profiles. The positive aspect, i.e. to practice a minority religion, be educated in the minority culture, speak the minority language, assert one's identity and minority lifestyle in public, etc, is presumably not relevant for group profiling. Third and finally, with group and minority rights, the groups are more or less formed before the group/minority rights are claimed; the members mostly want to be considered as part of the group. With group profiling, this is not the case, the group is formed by the profiling itself, the group members may have nothing more in common than that they have visited website X or have brown hair and they may not want to belong to the group they are profiled in. Their particular interests and feelings about being profiled may differ – some of them may object, others may be particularly keen on being categorized in a particular group and still others may be indifferent.

Reference to 'relational privacy' (which Bloustein (2003) somewhat unfortunately and misleadingly calls 'group privacy') or family privacy seems inapt as well. The idea is here that the family forms a unit, a group, and a privacy claim as a group might be attributed to it. The problem, however, which such a form of family privacy is of course that the group is involuntary for the child and that the problem of representation of the unit has historically been mainly 'solved' by letting the *pater familias* of the family represent it. The skepticism of most feminist writers has consequently been that privacy is used by males to conceal their wrongdoing inside the home, for example sexual or physical abuse. Privacy is often applied as a duty for women: they are only allowed to flourish inside the home, not in the public domain (Allen 1988; Elshtain 1991, 1995; MacKinnon 1989). It is thus in itself a highly contested concept and also the reason why for example the ECtHR does not allow families as such (as a unit) to complain about violations of their privacy. A somewhat better reference seems to be to the concept of future generations. There is an increasing branch in literature which argues for accepting rights of future generations (Gosseries 2008; Mayor Zaragoza 1996; Colb 2009; Shrader-frechette 2000; Gaba 1999; Davidson 2008), for example in relation to the right of future generations to a clean and healthy living environment, to the preservation of cultural heritage, etc. As is the case with group profiling, it essentially regards a claim of a not-yet-existent group. However, it must be admitted that the literature on future generations essentially focuses on the rights and obligations of current generations to protect the interests of future generations. Doing so, it does not facilitate a move towards accepting the rights of groups to claim a right to privacy to protect its own privacy interests. There is one further problem with this analogy, that is that future generations will exist and this is generally regarded as a positive thing, while the creation of group profiles is contested as such.

That is why, perhaps, the best analogy is the concept of wrongful life/wrongful birth.[53] This is the claim of children who argue that they should not have been born. Similarly, one could argue that groups should have a right to complain about their coming into existence. However, it is questionable whether this would provide a useful starting point, not only because the concept of wrongful life/wrongful birth is already quite contested, but also because it would be very difficult to specify which groups should not have come into existence. It is already difficult to determine which life is not worth living/should not have existed, though here, the effects on and consequences for a specific human being is quite clear. But with the existence of groups, this might prove far more difficult to determine. The use of group profiling seems necessary for the general functioning of governments, who, inter alia, need to differentiate between different socio-economic groups for their policies. Perhaps one could argue that group profiles on the basis of sensitive characteristics (such as race, religion or sexual orientation) should be regarded as problematic,

[53] See further: Gillon (1998), Robertson (1982), Ahuja (2011), Archard (2004), Picker (1995).

but even here, the determining factor seems not be the creation of the group profile as such, but what is done with it. If a group profile is used, for example, for positive discrimination, for example providing extra protection to Jewish communities in terms of security measures, the existence of a group profile might be regarded as positive. Moreover, the problem of group representation and participation, discussed earlier, remains. Finally, it should be kept in mind that the problem is not only for those who are included in a profile, but also that some are excluded from a profile (by virtue of character traits) and that groups of people (homeless people, illegal immigrants, etc.) are systematically excluded from databases.

Consequently, it might be argued that it is quite difficult and perhaps even impossible to find a suitable basis for building a new interpretation of the right to privacy. However, even besides the need for group privacy and group rights to other human rights, which has been discussed above, it seems that these counter-arguments are not decisive. This contribution started by distinguishing between two different approaches to privacy: perceiving it as a duty of the state not to use its powers arbitrarily and seeing it as a subjective right of the individual to protect his individual interests. It should be stressed that the idea of group rights is not contrary to the first approach. Rather, the authors of the European Convention on Human Rights explicitly recognized groups as a category of complainants. The argument that what is at stake with group profiling is a discrimination problem rather than a privacy problem only partially holds true. The core idea behind the Convention as a whole and the right to privacy in particular was preventing the *arbitrary* use of power, that is differentiating illegitimately between different groups in society or using power without any concrete reason at all. The fear of discrimination is thus inherent to all rights and freedoms granted under the ECHR. Finally, it should be stressed that the aim of the Convention was to address the structural and general abuse of power and this is precisely also the potential problem with big data analytics and its use through, among others, group profiling.

The second approach to privacy focuses on the protection of natural persons by granting them a subjective right to protect their personal interests, such as in relation to personal autonomy, human dignity and individual freedom. It seems that it is primarily this approach with which the idea of group rights might conflict. The question is whether groups should have a right to develop their identity and promote their interests as a group and perhaps more importantly, whether the group wants that. But if a group should want it, there seems no reason why it should not be able to invoke such a right. The only problems that remain are of practical nature. The problem of hierarchy and representation, the problem of determining who is a member of the group and who is not, the fact that groups and its membership may fluctuate at high speed, etc. These points all hold true, but they also seem to follow to a large extent from the exact legal framework that might be ready for revision. As discussed, the imparity between the level at which the infringements take place (group or general level) and the level at which the rights and remedies are granted (individual level) is not a problem specific to privacy, but also to the right to data protection, the freedom rights, the right not to be discriminated against, etc. Consequently, the developments here described challenge the human rights framework and perhaps even the legal framework as such. There are two equally attractive ways to move forward, the one not excluding the other. One is to change the fundamental premises of the human rights and legal framework, which is focused (at least in most Western liberal democracies) on the individual. Or, one could accept, as underlined already when discussing many of the issues following from big data processes, that these problems are in fact more aligned to ethical and political dilemmas than to the juridical domain. Perhaps, to accept group privacy is to move beyond the legal realm. In a way, this echoes the intentions of the drafters of the European Convention on Human Rights, who were very skeptical about the capacity of legal rules and legal remedies to address structural human rights problems. A narrow focus on the legal domain was, is and will remain insufficient to address many of the problems with human rights infringements in the age of big data, or any other age.

Bibliography

Ahuja, A. 2011. The case for wrongful life: The children encouraged to sue for being born. *New Scientist* 212(2836).

Allen, A. 1988. *Uneasy access: Privacy for women in a free society*. Totowa: Rowman and Littlefield.

Archard, D. 2004. Wrongful life. *Philosophy* 79(309): 403–420.

Bloustein, E.J. 2003. *Individual & group privacy*, 2nd ed. New Brunswick: Transaction Publishers.

Bygrave, L. 2002. *Data protection law: Approaching its rationale, logic and limits*. The Hague: Kluwer Law International.

Colb, S.F. 2009. To whom do we refer when we speak of obligations to "future generations"? Reproductive rights and the intergenerational community. *George Washington Law Review* 77(5–6): 1582–1619.

Davidson, M.D. 2008. Wrongful harm to future generations: the case of climate change. *Environmental Values* 17(4): 471–488.

Elshtain, J. 1991. *Public man, private woman: Women in social and political thought*. Princeton: Princeton University Press.

Elshtain, J. 1995. *Democracy on trial*. New York: Basic Books.

Gaba, J.M. 1999. Environmental ethics and our moral relationship to future generations: Future rights and present value. *Columbia Journal of Environmental Law* 24(2): 249.

Gillon, R. 1998. 'Wrongful life' claims. *Journal of Medical Ethics* 24(6): 363–364.

Gosseries, A. 2008. On future generations' future rights. *Journal Of Political Philosophy* 16(4): 446–474.

Lerner, N. 1991. *Group rights and discrimination in international law*. Dordrecht: Nijhoff.

MacKinnon, C. 1989. *Toward a feminist theory of the state*. Cambridge: Harvard University Press.

Mayor Zaragoza, F. 1996. *The rights of future generations*, UNESCO Courier, March, 1996.

Pariser, E. 2011. *The filter bubble: What the internet is hiding from you*. London: Viking.

Picker, E. 1995. *Schadensersatz für das unerwünschte eigene Leben: "Wrongful life"*. Tübingen: Mohr.

Raikka, J., and J. Aikk. 1996. *Do we need minority rights?: Conceptual issues*. The Hague: Nijhoff.

Robertson, G. 1982. Wrongful life. *Modern Law Review* 45(6): 697–701.

Shrader-frechette, K. 2000. Duties to future generations, proxy consent, intra- and intergenerational equity: The case of nuclear waste. *Risk Analysis* 20(6): 771–778.

Thaler, R., and C. Sunstein. 2009. *Nudge: Improving decisions about health, wealth, and happiness*. New York: Penquin Books.

Tomlinson, H. 2012. *Positive obligations under the European convention on human rights*, 2. http://bit.ly/17U9TDa, 2.

van der Sloot, B. 2014. Privacy in the post-NSA era: Time for a fundamental revision? *JIPITEC* 5(2014): 1.

van der Sloot, B. 2015. Do privacy and data protection rules apply to legal persons and should they? A proposal for a two-tiered system. *Computer Law & Security Review* 31(1): 26–45.

van der Sloot, B. 2016. Is the human rights framework still fit for the big data era? A discussion of the ECtHR's case law on privacy violations arising from surveillance activities. In *Data protection on the move*, Law, governance and technology series, vol. 24, ed. S. Gutwirth et al. Dordrecht: Springer.

van der Sloot, B. Forthcoming a. Privacy as a personality right: Why the EctHR's focus on ulterior interests might prove indispensable in the age of big data. *Utrecht Journal of International and European Law* 31: 25–50.

van der Sloot, B. Forthcoming b. Privacy as virtue: Searching for a new privacy paradigm in the age of Big Data. In *Räume und Kulturen des Privaten*. Heidelberg: Springer.

van Dijk, P., F. van Hoof, A. van Rijk, and L. Zwaak (eds.). 2006. *Theory and practice of the European convention on human rights*. Antwerpen: Intersentia.

Chapter 12
Conclusion: What Do We Know About Group Privacy?

Linnet Taylor, Bart van der Sloot, and Luciano Floridi

Abstract This chapter draws together the conclusions of the book as a whole, namely that the available typologies of group privacy, such as collective action lawsuits or the articulation of the rights of political or activist groups, are intuitively insufficient to address the landscape emerging from the new data analytic technologies. The book has demonstrated that there are multiple and often divergent perspectives on what a group is and how it should be addressed with regard to privacy, but this divergence is also an important tool for understanding which elements of the problem can be addressed using current legal and conceptual tools and which will require new approaches. The chapter outlines the authors' contributions to the typology of group privacy concerns, then identifies the gaps and limitations that arise from a group perspective on privacy, and the conceptual and practical implications of taking the group level into account. Finally, it suggests ways forward for future research, discussion and action.

Keywords Typology • Privacy terminology • Social hierarchy • Relationalism • Discrimination • Class action • Group rights • Data protection

L. Taylor (✉) • B. van der Sloot
Tilburg Institute for Law, Technology and Society, Tilburg University
90153, Warandelaan 2, 5000 LE Tilburg, The Netherlands
e-mail: l.e.m.taylor@uvt.nl; B.vdrSloot@uvt.nl

L. Floridi
Oxford Internet Institute, University of Oxford, 1 St Giles, OX1 3JS, Oxford, UK
e-mail: luciano.floridi@oii.ox.ac.uk

L. Taylor et al. (eds.), *Group Privacy*, Philosophical Studies Series 126,
DOI 10.1007/978-3-319-46608-8_12

This book has demonstrated that the available typologies of group privacy, such as collective action lawsuits or the articulation of the rights of political or activist groups, are intuitively insufficient to address the landscape emerging from the new data analytic technologies. The ways in which data technologies can be used silently not only to group but also to modulate behaviour has profound implications for people's ability to interrogate and understand those technologies. What are the lessons to be learned from the cases in this book for debates about privacy, and particularly for the public's ability to engage in those debates? How should accountability be conceptualised in an era where almost everyone is constantly being grouped and regrouped, unaware, by data analytics?

We have addressed the challenge today's data technologies pose to privacy as a multidisciplinary question, an approach that highlights both points of agreement and issues of divergence that will, we hope, be useful in moving the issue of group privacy forward both conceptually and practically. Both are necessary in light of the ways in which data analytics are currently changing the way groups are conceptualised. Every day, algorithms are used to produce 'calculated publics' (Gillespie 2014) through which people are grouped, influenced, nudged or herded, unaware, towards certain kinds of behaviour or governability. Contributors to this volume from the fields of sociology, law and computer science have discussed the use of big data to sort and categorise in the fields of health (de Hert and Hallinan), human rights (Raymond), development (Taylor; Kammourieh et al.), and how it acts as a way to group, bind and use human capacity and understanding (O'Hara and Robertson). Beyond this, research is showing the utility of big data for guiding decisionmaking on issues ranging from urban planning (Bettencourt 2013) to national security (Lyon 2014), health (Wesolowski et al. 2012) and disaster response (Bengtsson et al. 2011).

Given these myriad ways in which the impacts of data analytics may play out, the initial contribution of this volume is a typology of group privacy challenges. The authors of the various chapters make clear that there are multiple and often divergent perspectives on what a group is and how it should be addressed with regard to privacy, but this divergence is also an important tool for understanding which elements of the problem can be addressed using current legal and conceptual tools and which will require new approaches. In this concluding chapter we will begin by outlining the authors' contributions to draw together this typology of group privacy concerns, and will then discuss what these types can tell us about the possible place of group privacy in scholarship, policy and law. Next we will identify the gaps and limitations that arise from a group perspective on privacy, and the conceptual and practical implications of taking the group level into account. Finally, we will suggest some ways forward for future research, discussion and action based on the findings in this volume.

12.1 Types of Group Privacy Challenge

The terminology used by the authors of this book differs to a large extent. Terms used include ‘unit’, ‘group’, ‘aggregate’, ‘cluster’, ‘class’, ‘collective’, ‘network’, ‘category’, ‘artificial person’, etc. No unified terminology is used by the authors, nor does one clear proposal for understanding group privacy emerge. Given the wide variety of approaches, however, it is possible to categorise the different meanings, understandings and applications of ‘group privacy’. There are a number of factors that can be taken into account when describing groups and group privacy, of which the most important seem to be: are the groups real or artificial, are they self-proclaimed or framed, are they self-aware or not, are they stable or fluid and are they hierarchical or egalitarian?

Real or Fiction Floridi discusses in this book the tension between philosophers who have claimed that groups or categories are discovered and those who have argued that they are invented. Platonists have traditionally argued that the group precedes the particular, that the category ‘horse’ logically precedes the existence of specific horses. On the other side of the spectrum are those that have argued that only units are natural, and that the individual is the natural building block from which categories or groups can be designed. In the latter interpretation, the group and its interests cannot and should exceed the unit and the individual interest. As Pagallo has forcefully shown, the legal domain is quite aware of this dichotomy and has traditionally firmly embraced the latter interpretation of groups. The individual human is called the natural person and legal persons or groups are commonly referred to as fictional persons. This has also had an impact on the extent to which the different categories are protected. Although fictional persons have been part of the legal domain from the beginning, as Van der Sloot has shown, fictional persons are not, or to a far lesser extent, protected under traditional human rights frameworks and are usually not said to have a right to privacy.

Importantly, Floridi takes a middle ground and argues that groups are designed by the level of abstraction at which a specific analysis of a social system is developed. Their design is therefore justified insofar as the purpose, guiding the choice of the level of abstraction, is justified. He argues that the naturalness of a grouping is just a function of the intuitiveness of a level of abstraction, that is, it is epistemological, not ontological. Referring to salad, tomatoes and potatoes as a group called food seems something as observer-independent and objective as possible, but this is only because we assume our own interests as organisms and eaters as the natural, intuitive, and relevant level of abstraction. To a tiger, they would all look as unrelated and inedible as grass and leaves do to us. Accepting that our knowledge of the world is obtained through different levels of abstractions is not to say that anything goes, and that the only alternative to nominalism and realism is some kind of untenable relativism. It is to say that absolute questions asked in a logical space lacking any references and orientation (interest, purpose) are an absolute mess, and that *relationalism* is a better alternative. Using the previous example, asking whether something is food means adopting the right level of abstraction at which it makes sense to ask whether a specific substance can be a nutrient for a specific organism. Food is a relational (not a relative) concept: it takes a level of abstraction with two relata to define it, yet not every level of abstraction is correct and some level of abstractions will be more correct than others. According to Floridi, this removes the objection that groups cannot have a right to privacy because groups are mere artefacts (there are no groups, only individual persons to which groups are ultimately reducible) or that, even if there are groups, it is too difficult to deal with them.

Self-Proclaimed or Framed There is an important difference between the types of groups brought to light by new data technologies and those we have traditionally been acquainted with. Although in the past some groupings were formed by others, for example in order to lay down discriminatory policies, and the legal domain granted these groups a right against discrimination, most groups were self-proclaimed. Reference can be made to Manchester United fans or other sport fans setting up a club, religious groups founding religious orders or sects, music lovers, bikers, or any self-proclaimed groups that are united in some way or another. Membership is usually self-initiated and voluntary. Not only does the legal realm grant groups a negative right against being grouped and discriminated against by others, certain self-proclaimed minority groups have also been attributed positive rights. Minority rights grant groups claims, for instance, to educate themselves in their own language, to engage in a form of self-government, or to practice their cultural traditions and customs[1] These people want to be seen as part of that group and want to develop their group identity.

Most authors in this book, however, have referred to non-self-proclaimed groups, either designed or discovered. This seems to be the general concern in the big data era. The focus is on the data controller, the one having access to the data, analysing them and using them for policy purposes. It is the data controller that usually has the power to categorize and form groups. Kammourieh et al. refer to four types of categories. First, data analytics can help to find out new things about pre-identified groups. Although the group might have been pre-defined, we now have the opportunity to infer new information from data about it without having any pre-defined hypothesis in place. Pre-identified groups can be self-proclaimed, but need of course not be. Second, we might come to identify previously non-apparent groups on the basis of certain pre-defined parameters. Third, without defining any parameters or characteristics up front, we might discover groups through new analytical approaches. This can lead to the identification of new groups on the basis of previously unknown sets of characteristics. Fourth and lastly, while using such analytical processes, there will be an increasing risk that we remain unaware of the discovery of new groups, even as the claims resulting from the analysis might affect or harm them. The group thus remains latent. This is possible in two scenarios: either the group has been identified within the data mining process itself but has not become apparent to observers; or a group classification has been enforced through the

[1] http://www.un.org/esa/socdev/unpfii/documents/DRIPS_en.pdf

analytical process by the choice of certain data which are non-representative or biased in some way. Consequently, most groups in the big data era seem non-self-proclaimed. This is important, among other reasons, because self-proclaimed groups often want to be seen as a group and the units want to belong to it, while this is not necessarily the case with enforced groups.

Self-Aware or Not First, we can identify a category of arguments dealing with groupings that are selfaware and purposefully formed, and where privacy challenges are on a categorical level. This includes journalists, political groups, rights activists, or a grouping of multiple organisations as in the case of Civil society vs. Intelligence services, discussed by Eijkman in relation to bulk interception of communications by governments. This is also the type of grouping Pagallo deals with – groups that can claim rights that may conflict with the rights of their members.

Second, our contributors identify groupings that are not selfaware, possibly the most extreme example of which is what de Hert and Hallinan refer to as a genetic category – those who share a certain genetic architecture but are not aware of it, and therefore have no understanding of the implications of analysis of their genetic code. The only way the law can currently address these people, de Hert and Hallinan contend, would be as 'incapax', or incapable of intent. Taking the problem further, Taylor and Raymond's contributions identify the problem of data analytic categorisation and resulting intervention where the group in question is not only unaware of a collective identity, but cannot be individually identified even by the analyst. With this kind of group, the only way to address privacy concerns is through ethical decision-making at the data analysis stage, rather than at the point of collection or use.

Finally, O'Hara and Robertson identify a middle ground of emergent groupings where people collaborate through technology towards a particular aim and are therefore connected functionally, but not socially. In this case, the ability to keep certain aspects of the group's information secret is essential to the emergence of the group itself – so this might be termed privacy through grouping rather than the privacy of groups.

Stable or Fluid There is also a difference between stable and non-stable groups. Stable groups can be further divided into self-proclaimed and biological groups. Self-proclaimed groups are relatively stable because they want to be seen and possibly treated as groups. Of course, the defining factors of a group and who belongs to it may vary slightly over time. Likewise, groups formed on the basis of biological criteria are relatively stable, because biological criteria such as race, gender and DNA sequence, to name but a few criteria, though not absolute, are relatively stable. Many of the traditional anti-discrimination provisions also refer explicitly to these types of factors. For example, Article 14 of the European Convention on Human Rights refers to sex, race and colour, among other characteristics. In a similar vein, Hallinan and De Hert refer to genetic groups. Although genes may change over time due to natural mutation, they are relatively stable.

The fact is, however, that in the big data era, groups are increasingly fluid, not only through their changing membership, but also because of the changing criteria for the group itself. A group the criteria for grouping people and the membership of a group might change in a split second. The purpose for which the group is designed may also change from day to day to adapt to new insights gained from data analytics, and groups may be formed and dissolved through the push of a button. Because big data analytics make it so easy and convenient to form groups, their number (at least of which the data controller is aware) has sky-rocketed over the last few years. Consequently, one person may easily belong to a thousand groups or more at a given moment in time. This is important because in the ethical, legal and social spheres claims, rights and interests are mostly attributed to relatively stable groups. From a pragmatic point of view, it seems undesirable and impossible to grant groups rights if the groups, the criteria for grouping people and their membership could change in a split second. Moreover, if a person is a member of more than a thousand groups at a given moment in time, of which he or she is barely aware. Consequently, it would be an almost Sisyphean task to grant all such groups a right to protect their interests, and therefore even if this is important, another way must be found.

Hierarchical or Egalitarian One thing that makes granting groups rights even more problematic is that groups, as opposed to legal persons, traditionally have no or a very limited hierarchy. In the legal domain, groups must be distinguished from legal persons (for example, corporations). Legal persons have a legal persona because they are relatively stable, because they usually have a fixed identity and because they have a hierarchy within their organisation. As Mantelero points out, the problem with many fluid and non-hierarchical groups is that the interests of the group members might conflict. When categorised as a 'male conservative sports-fan', one member belonging to the group might be quite happy because the content of, for example, a news website is adjusted to suit his profile, while another might be neutral to being profiled as such, another might object because she is falsely categorised in this group, and still another person might be correctly categorised, but still object because he does not want to receive personalised content.

As Van der Sloot argues, the legal system in general tends to stimulate groups to obtain a legal status and a form of hierarchy. This ensures that the government, the judge or any other organisation knows who is the representative of a certain group or minority. The leader, boss or board can then be seen as the legal representative. They can also decide what is in the interest of the legal person, which ensures that one unit has one (proclaimed) interest. But when a group is not a legal person, this point is not resolved. Suppose that a specific group or community was stigmatised by a law or policy and wanted to challenge its position before the court: should the whole community submit a complaint, or should each and every member join as an individual claimant to this complaint, or should it create a legal organisation in order to represent it? Rather quickly, the tendency is to choose either a form in which individual complaints are bundled and aggregated, a form in which one individual is said to represent the whole group or a form in which a legal organisation has the task of legal representation. Thus, the tendency is to move to collective or corporate rights, but the question is whether a group can itself invoke a right to privacy.

12.2 Types of Interests to Be Protected

Besides the difference in types of groups, it is important to distinguish between the different interests that could potentially be protected through group privacy. As has been referred to several times in this book, Bloustein (1978), one of the first to coin the concept of 'group privacy', used it not in the sense of a group interest to be protected, but the interest of individuals in belonging to a group, forming a group or protecting their identity as part of a group identity. This is commonly referred to as family privacy or relational privacy. Relational privacy has been commonly accepted in most traditional social, ethical and juridical privacy paradigms, which focus on the individual claiming the right to protect an individual interest – that is, the interest in standing in relation to others. For example, Article 8 of the European Convention on Human Rights, besides protecting the right to private life, home and communications, also protects family life. Moreover, the European Court of Human Rights has from very early on stressed that the right to privacy includes 'the right to establish and to develop relationships with other human beings, especially in the emotional field for the development and fulfillment of one's own personality.'[2]

This book has tried to move beyond these more traditional interests, however, and the types of interests linked to group privacy here are quite varied. Some authors see as a focal point the fact that central to big data processes is no longer personal identifying information, but rather group information or data about units or categories. This is why it might be useful to broaden the term 'personal data' so that it encompasses not only data about natural persons or about legal persons, but also about groups. Other authors have stressed that in any case, processing anonymous data and metadata can now be just as problematic as processing personally identifying information since, among other reasons, the analysis of metadata can give both a very detailed picture of the content of the communication and of a person's private life. Likewise, the value of general data is growing rapidly because even public and non-sensitive data can be used to adopt and apply far-reaching policies and measures and even to implicitly discriminate against certain groups (also known as redlining). In addition, it is important to point out that even although data may be aggregated, anonymised or encrypted, reversing de-identification is increasingly easy in the big data era. That is why several authors have suggested that it is important to critically assess aggregated and anonymised data, that is, to regulate group profiles and statistical data alongside personal data.

Others have pointed to the fact that both the positive and negative effects of data analytics increasingly have an on impact groups rather than individuals. Although the individual obviously feels the effects, they are affected due to group membership.

[2] X. v. Iceland.

Thus, the group 'males with low education, living in a rural area' might be discriminated against. Although the current ethical and legal regime protects Joe, who is a male with low education living in a rural area, and safeguards him from discrimination, this is increasingly missing the point. It is problematic that the group as such is created and treated negatively, but it is not as such problematic that Joe belongs to a certain group or is categorised as such. It is also important to note network effects: that if one person is targeted, this might have an effect on the people around them. If a person is denied a loan, for example, because he is male, with a low education and living in a rural area, this might also impact his wife, his children and potentially, other family members as well.

Then there are the full-fledged group interests, four of which might be distinguished on the basis of the contributions to this book: a negative, a positive, a constitutive and a discontinuing interest. The negative group interest is that of not being discriminated against by others. This is the classic idea contained in discrimination law, namely that states, institutions or natural persons may in principle not (ab)use group identities based on race, religion, gender or political or sexual preference to treat people in a negative manner. The positive interest of a group is typically associated with minority rights, to which reference has already been made. Such an approach might include the right of minorities or indigenous people to education in their own language, to self-government and to practice their cultural traditions and customs. The constitutive interest protects the very basis on which the group is formed. Floridi, for example, has referred to a group wanting to hold a private funeral or a sect wanting to hold secret religious meetings. The privateness of their actions are constitutive for the forming of these groups. Similarly, Van der Sloot refers to the idea that secrecy is a necessity for the functioning of groups such as lawyers (the secrecy between lawyer and client), doctors (the secrecy between doctor and patient) and for the democratic process (the secrecy of ballot). Without, for example, the secrecy between doctor and patient being guaranteed, people will simply not go to the doctor and when they do, will disclose less information than they would if secrecy were protected. Finally, there is the discontinuing interest, namely the fact that especially in big data processes, groups are formed and people are categorised without their knowledge and against their will. Consequently, their main interest lies in the group not being formed or their not being categorised as members. As such, it differs from a negative interest, which does not address the problem of the group being formed, but the fact that negative consequences are attached to the existence of the group.

Finally, a number of authors have focused on the protection of the general interest, for which either individuals, groups or legal persons may be responsible. For example, Eijkman has referred to strategic litigation in the common interest, such as through class actions. Class actions are often initiated by civil society organisations, such as Amnesty International, Big Brother Watch and Privacy First. For example, class actions are put forward in many countries and with European courts such as the European Court of Human Rights on mass surveillance activities by states. Here individuals, groups and legal persons may protect a general, societal interest and/or everyone's interest not being subjected to mass surveillance. This differs from traditional cases revolving around discrimination law, in which, for example, gay people or gypsies are discriminated against – in such cases, individuals may go to court because they are negatively affected as a member of a group, but groups or legal persons may in principle not do so on their behalf. Consequently, in the latter type of case it is the collective interest, as stated by Mantelero, that is protected, i.e. the aggregate of the interests of several individuals. With mass surveillance cases, however, many courts, including the European Court of Human Rights, have accepted claims from non-individuals protecting the general interest.

12.3 The Implications of a New Concept of Group Privacy

It is clear that the idea of group privacy has little to no legal traction at present. The legal arguments in this book demonstrate why it will be hard for it to gain that traction. Yet new data technologies such as those that enable the bulk interception and analysis of communications, the actionable categorisation of people without their knowledge, and invisible monitoring of groups' movements and activities do raise questions that go beyond the level of individual privacy harms, and that are experienced by groups, both selfaware and unknowing. Furthermore, new challenges are making it likely that legal instruments will be called into play to defend group rights with regard to these data technologies, as is occurring with the EU data protection regulation which provides the potential to protect genetic categories that are not selfaware (de Hert and Hallinan).

Given that group rights may emerge gradually in response to particular challenges, the main obstacle to a workable concept of group privacy may in fact be institutional – we may lack the institutional configurations, the right actors may not currently be empowered to act and regulate, and the populations at risk of harm from the new data technologies may not have either the informational tools to become aware of the problem or the legal or rights instruments to seek redress. For example, de Hert and Hallinan posit that in order for the GDPR to come into play with regard to genetic discrimination based on categories people are unaware of belonging to, an institution such as a genetic rights NGO may have to step in on behalf of those who are unaware their rights are at risk. Similarly, in Taylor's example of resistance to technological visibilities imposed by developmental states on marginalised people, the imposition of visibility goes hand in hand with a lack of institutions capable of representing the rights and preferences of data subjects.

How should the boundaries of analytics and hence of group definition be drawn? As de Hert and Hallinan point out, data processing actions regarding a particular genetic group may create conclusions with implications for other genetic groups, or indeed everybody. There are also gaps with regard to basic principles of data protection. In light of big data analytics some principles and concepts essential to data protection become unstable and need to be revisited, for example the Fair Information Practice Principles, the concept of data's accuracy with regard to groups, and the assumption that attention should be focused on the legitimacy of data collection and its end uses. In fact, however, many of the contributors to this volume demonstrate that the most serious risks may occur at the stage of data processing, since it is often hard to predict the precise purpose of processing where big data is involved, and therefore ethical and practical decisions about how data should be used and what it should bring to light may now occur at various points in the analytical process.

Tarleton Gillespie (Gillespie 2014) has argued that '[w]e don't have a sufficient vocabulary for assessing the algorithmic intervention.' The problems outlined in this book sugges that it is time to create this vocabulary, if only because doing will also provide new tools with which to think critically about how to manage the implications of calculated publics. Such a shift would also, however, result in a more diffuse focus for data protection instruments. This might include a move towards formulating ethical assessment tools that are more probabilistic than the Privacy Impact Assessments currently in use as compliance tools by data controllers, and that focus on how risks may be heightened by certain analytical decisions and processes along the whole sequence of data collection, analysis and use. Such a move beyond compliance to a broader understanding of the risks of data analysis would be a paradigm shift for all actors in the data market, and would also create challenges for regulators who rely on clear rules with regard to what is permissible.

However, regulators are only one group responsible for influencing how data is used and shared. Much of the data market currently operates on a basis of corporate selfregulation since state regulators lack the capacity to monitor and respond to the current massive scale of data collection, sharing and use. Moreover, the field of data analytics extends beyond governments and the private sector to a very broad range of actors including multilateral institutions, almost all branches of academic research and scientific knowledge, and philanthropic institutions. At this point in the big data field's evolution norms are arguably just as important as rules, and normsetting is one area where debates about group privacy could usefully contribute.

The idea of group privacy also casts a new light on the way that privacy interacts with data protection. As Pagallo (this volume) says: 'That which may make sense in privacy law does not necessarily fit the field of data protection.' The idea of data analytics' effects on the group level brings us to think about regulating, or at least providing mechanisms for considering, the collection of data and the process of analysis, rather than the results and their use. But it also suggests that we need to find ways to protect groups who cannot communicate, and who are unaware of a potential challenge to their rights – for instance, by allowing data protection authorities to investigate without a complaint being filed, and in advance of processing.

A group perspective on privacy, importantly, also moves the focus from processes of consumption to those of citizenship and accountability. It does so because it exposes the difficulties inherent in demands that people manage their own data and privacy, and counters the convenient fiction that people can be made responsible for auditing and managing flows of data about themselves. As senior Microsoft official Craig Mundie has said, 'today, there is simply so much data being collected, in so many ways, that it is practically impossible to give people a meaningful way to keep track of all the information about them that exists out there, much less to consent to its collection in the first place' (Mundie 2014). If we eschew the idea that people can manage their own privacy, however, we are forced to consider the relationship between the data economy, citizenship and the social contract, and to ask whether adding a grouplevel consideration to privacy rights and instruments would potentially make questions of automated sorting something that should be managed through democratic debate and political accountability, rather than fair business practice.

Although it is clearly useful as a tool for thinking about how privacy needs to develop in the future, there are also negatives to adding a group perspective to current conceptualisations of privacy. As Pagallo points out, there are important instances where we may disempower individuals by taking the group into account, privileging the rights of the many over the voices of the marginalised or vulnerable. This is the argument feminist scholars have long offered against an absolute right to privacy in the domestic sphere (Gelles and Straus 1988). We should also be careful of advocating a group privacy approach as a stop-gap for the fundamental individual right to privacy in situations where that right is not fully conceptualised, or is difficult to claim or enforce. A group perspective is a risky rather than a valuable tool if it becomes a way for authorities to claim that they have respected rights by evaluating the risks of data analytics only on the population level. The chapters by Taylor and Raymond suggest that some of the best examples arguing for a concept of group privacy do indeed come from places where rights are already being abused or are more difficult to enforce – making it clear why we should consider a group level of privacy as an enhancement and safeguard for the individual right to privacy, rather than as a potential substitute for it.

12.4 Ways Forward

This volume demonstrates that it is not only the blackboxing of big data analytics, but also the remoteness and decontextualisation of those analytics and the decision-making based on them that make it practically impossible for individuals to perceive or contest either. Individuals may participate willingly in social machines (O'Hara this volume) built on stable privacy provisions, but as a whole, the authors who contribute to this book suggest that the 'networked public' (Boyd 2010; Mizuko 2008) is not a solution to the problems raised by data analytics since the calculated publics of big data have little agency regarding how they are constituted, and are therefore in danger of becoming both disarticulated and disenfranchised.

Perhaps more appropriate is Dewey's image of the public as a body constituted by 'the indirect consequences of transactions'(Dewey 1991[1927], 15–16). For Dewey, it was the role of the body politic as a whole to engage with the consequences of groupings created through policy, since the individuals thus grouped were not able to effectively monitor or resist. In order to find the kind of agency that can respond to the operations of today's data analytics, we should ask how the discussion of rights with regard to data technologies has somehow moved out of the democratic and into the technocratic sphere. In order to understand how people may regain rights over the way they are categorised and acted upon through data, we may need to move from 'privacy by design' to 'privacy by accountability'.

For group privacy to evolve into a concept that can play a role in privacy and data protection rights, the arguments presented by the authors in this volume suggest that we may need to work from specific problems rather than broadening the conceptualisation of privacy *per se*. Group privacy can be seen as an important complement to individual privacy, provided that its ability to subsume the individual is kept in check. If we start from the problems laid out in this book – genetic profiling, policy intervention for behaviour change or security, mass surveillance via sensing technologies – we can locate just some areas where it is clear that individual privacy does not go far enough.

A clue to how this concept may develop can also be found in the composition of the debate presented here: the often-conflicting perspectives show some areas where it is necessary to step back, and many where it is possible to push forward. If group privacy emerges gradually in response to specific challenges it will be hard for it to eclipse individual rights, and it will be modulated by the inevitable accompanying debate and legal challenges. Taking a case-specific approach, however, also allows nuanced commentary from domain experts, and may lead (as this book demonstrates) to conclusions that are very different from those of a purely legal debate. Multiple paths and multidisciplinarity may therefore prove more productive than a search for agreement, and the search for *a* way forward may be more important than a search for *the* way forward. This book is a contribution to such an incremental and multidisicplinary approach: we hope it will be a productive one.

Bibliography

Bengtsson, L., X. Lu, A. Thorson, R. Garfield, and J. von Schreeb. 2011. Improved response to disasters and outbreaks by tracking population movements with mobile phone network data: A post-earthquake geospatial study in haiti. *PLoS Medicine* 8(8): e1001083 1–9. http://doi.org/10.1371/journal.pmed.1001083

Bettencourt, L.M.A. 2013. *The uses of big data in cities* (No. SFI working paper: 2013-09-029).

Bloustein, E.J. 1978. *Individual and group privacy*. New Brunswick: Transaction Publishers.

Boyd, D. 2010. Social network sites as networked publics: Affordances, dynamics, and implications. In *Networked self: Identity, community, and culture on social network sites*, ed. Z. Papacharissi, 39–58. New York: Routledge.

Dewey, J. 1927. *The public and its problems*. Athens: Swallow Press/Ohio University Press.

Gelles, R.J., and M.A. Straus. 1988. *Intimate violence: The causes and consequences of abuse in the American family*. New York: Simon & Schuster.

Gillespie, T. 2014. The relevance of algorithms. In *Media technologies: Essays on communication, materiality, and society*, ed. T. Gillespie, P. Boczkowski, and K. Foot, 167–194. Cambridge, MA: MIT Press.

Lyon, D. 2014. Surveillance, snowden, and big data: Capacities, consequences, critique. *Big Data & Society* 1(2). http://doi.org/10.1177/2053951714541861.

Mizuko, I. 2008. Introduction. In *Networked publics*, ed. K. Varnelis, 1–14. Cambridge, MA: MIT Press.

Mundie, C. 2014, March/April. Privacy pragmatism. *Foreign Affairs*. Retrieved from http://www.foreignaffairs.com/articles/140741/craig-mundie/privacy-pragmatism

Wesolowski, A., N. Eagle, A.J. Tatem, D.L Smith, A.M. Noor, R.W. Snow, and C.O. Buckee 2012. Quantifying the impact of human mobility on malaria. *Science*:267–270. http://doi.org/10.1126/science.1223467

Printed by Printforce, the Netherlands